Brain, Mind and Consciousness

Petr Bob

Brain, Mind and Consciousness

Advances in Neuroscience Research

Springer

Petr Bob
Center for Neuropsychiatric Research of Traumatic Stress
and Department of Psychiatry
Charles University
Prague, Czech Republic
petrbob@netscape.net

ISBN 978-1-4614-0435-4 e-ISBN 978-1-4614-0436-1
DOI 10.1007/978-1-4614-0436-1
Springer New York Dordrecht Heidelberg London

Library of Congress Control Number: 2011936223

Printed on acid-free paper

Springer is part of Springer Science+Business Media (www.springer.com)

Preface

More than three centuries ago, Rene Descartes looked for "the seat of the soul" within the brain that could integrate *res cogitans*, representing the inner world, with *res extensa* – the outside world. Descartes thought that this special place is involved in sensation, imagination, memory, and the causation of bodily movements, and described the mind as an extracorporeal entity. In this theoretical concept Descartes intuitively anticipated the so-called binding problem of consciousness, which means that there is a part of the nervous system that integrates and transforms neural activity into reportable subjective experiences in the hypothetical center later called the Cartesian theater.

In agreement with this Cartesian concept there is evidence that certain parts of the brain are more essential for consciousness than others and may represent local integrative centers. On the other hand, there is a conceptual approach to consciousness that suggests that consciousness (instead of a single central place – "Cartesian theater") might be related to the binding of various events represented by groups of synchronized excited neurons that are located at different parts of the brain without unifying spatial convergence. This neural activity occurs synchronously across brain regions and likely underlies the integration of diverse brain activities. Together these findings indicate that a solution of the binding problem might reside within the fundamental problem of consciousness in modern neuroscience.

The subtitle of the well-known book *The Astonishing Hypothesis* and *The Scientific Search for the Soul.* In this book, Crick argued that the traditional Cartesian concept of the soul as a nonmaterial being must be replaced by a scientific understanding of how the brain produces the mind. Although this problem is still unresolved, there is a predominant opinion that consciousness emerges from a dynamical nucleus of persisting reverberation and interactions of neural groups. Other approaches to the binding problem also include non-conventional hypotheses related to various physical theories, such as complexity and chaos theory, quantum physics, and the theory of relativity.

Although we do not know how the nervous system integrates distributed neural activities and creates subjective experience, there is evidence that disturbed neural interactions producing patterns of temporal disorganization, decreased functional connectivity and global distribution of information, may influence the mind. Additionally, there is evidence that deficits in functional connectivity and distribution of information may underlie specific perceptual and cognitive states related to the disintegration of consciousness that may occur in various neuropsychiatric disorders.

On the other hand, disturbances in neural synchronization and coherence likely present a basis for discrete mental states that through differences between them enable recognition and awareness of the external and internal world. This form of internal disunity of the brain likely presents basic code that defines relative differences and enables their recognition in mental and physical space. Through this process specific observers may define reality and create observer-specific neurogeometry of the space and time.

To the basic assumption that causal connections exist between the brain and mind, recent neuroscientific evidence has added another – that the mental state may significantly influence the brain and body on various functional levels. In addition, this interaction between mind and brain enables the laws of nature to be discovered and the external world to be understood, likely through the rules that integrate the basic nature of the mind and the physical world. Because we can compare all differences only with respect to the unity, there is something in us that is immutable; as C.G. Jung said: "They are only an illusion, time and space, and so in a certain part of our psyche time does not exist at all." This unknown likely enables that we want to know.

Petr Bob

Contents

Chapter 1
Brain Structures and Consciousness

This chapter introduces the problem of localization of mental functions within the brain. According to recent evidence mental functions and consciousness are related to specific brain structures, but at the same time there is evidence that mental functions and consciousness are related to the binding of diffuse and synchronized neural activities. Recent findings strongly suggest that the neural binding cannot be simply explained paradigm suggesting localization of the mental functions, thus necessitating substantial revision of the Cartesian concept of the brain and localization of consciousness.

There is historical scientific evidence that certain mental functions can be understood within the concept of localization in the brain. Historically, it has been suggested that Paul Broca was the first to outline the concept of localization in the 1860s, when he studied brain deficits in aphasic patients and found the brain center for speech production to be located in the ventral posterior region of the frontal lobes, now known as Broca's area. Shortly after Paul Broca's work came to light, Carl Wernicke published his findings and found that not all language deficits are the result of damage to Broca's area; damage to the left posterior superior temporal gyrus now referred to as Wernicke's area, results in deficits in language comprehension. Other historically significant findings on the localization of the brain and mental functions were reported in the 1870s by Gustav Fritsch and Eduard Hitzig, who studied the electric localization of brain functions in experimental animals and found that the electrical stimulation of different brain areas causes involuntary muscular contractions of specific parts of the body (Finger 1994).

Other distinguished findings on the localization of mental functions were reported in the 1940s by neurosurgeon Wilder Penfield, who found that stimulation of the motor cortex of the left or right hemisphere produces effects of muscle contraction on the opposite side. Penfield's findings represented the first evidence of the brain's laterality of mental functions, which was later significantly developed in the 1960s by Michael Gazzaniga and Roger Sperry in their research into patients who had undergone corpus callosotomy ("split-brain patients") owing to severe epilepsy. Because the corpus callosum connects both brain hemispheres, the partial cutting of the corpus callosum reduces interhemispheric communication and the

P. Bob, *Brain, Mind and Consciousness: Advances in Neuroscience Research*,
DOI 10.1007/978-1-4614-0436-1_1, © Springer Science+Business Media, LLC 2011

spread of epileptic activity, and may sometimes be used for epilepsy treatment to reduce seizures. In these patients Gazzaniga and Sperry reported very interesting findings on the brain lateralization of various mental functions and perceptual processes in the left and right hemispheres (Kandel et al. 2000).

Enormous progress in the study of brain localization of the mental functions led to research using neuroimaging techniques, in particular functional magnetic resonance imaging (fMRI) and positron emission tomography (PET), whose importance lies in their sensitivity for revealing subcortical brain structures with high spatial resolution. Experimental research using these methods provides evidence that an increase in brain activity during a specific cognitive task, for example, during human speech, is not strictly localized to prominent brain areas, but is distributed to various parts of the brain. These findings strongly suggest that the neural correlate of mental functions and consciousness might not be strictly localized to one part of the brain and that highly diffuse activities are coupled with localized processes in certain distinct areas of the brain. In this context there is growing evidence that the neural correlate of consciousness likely represents a coherent neural process that connects diffuse brain activities to make a coherent whole (Crick and Koch 1992; John 2002), which cannot be simply explained within the paradigm, suggesting localization of the mental functions.

1.1 "Functional Neuroanatomy" of Consciousness

Some ideas about the localization of consciousness in the brain have been the focus of scientific thought from the beginnings of modern science. More than three centuries ago Rene Descartes described the problem of brain localization of consciousness and saw the pineal gland as "the seat of the soul." In his book he wrote that "… although the soul is joined to the whole body there is a certain part where it exercises its functions more than all the others" (Passions of the Soul, p. 31). Descartes thought that when we sense only one image with two eyes, only one sound with two ears or only one object by two hands, the sensations from two sources must be fused somewhere (Barrera-Mera and Barrera-Calva 1998; Smith 1998). Descartes intuitively postulated that this information is fused and governed by a mechanism in the pineal gland. He believed that the pineal gland is involved in sensation, imagination, memory, and the causation of bodily movements, and described the mind as an extracorporeal entity that is expressed through the pineal gland (Barrera-Mera and Barrera-Calva 1998; Smith 1998). Descartes intuitively anticipated the so-called "binding problem" of consciousness, i.e., where and how the information from various sensory modalities is integrated into the whole. The neural correlate of consciousness is thus a part of the nervous system that transforms neural activity into reportable subjective experiences (Fig. 1.1). The major hypothesis is that the neural correlate of consciousness can compare and bind activity patterns only if they arrive simultaneously at the neural correlate of conscious experience (van de Grind 2002).

Fig. 1.1 Rene Descartes

Consciousness combines the present multimodal sensory information with relevant elements of the past and creates spatiotemporal memory. Information from each modality is continuously distributed into distinct features and locally processed in different, relatively specialized, brain regions and globally integrated by interactions among these regions. Information is thus represented by integration through levels of synchronization within neuronal populations and of coherence among multiple brain regions that facilitate large-scale integration, or "binding" (John 2002; Singer 2001; Crick and Koch 2003; Zeman 2001; Bob 2009).

Recent hypotheses dealing with specific brain structures that could play a key role in the existence of the neural correlate of consciousness emphasized a specific influence of the anterior cingulate (Cotterill 1995), hippocampus (Gray 1995) or intralaminar nuclei of the thalamus (Bogen 1995), but later, Weiskrantz (1998) proposed that a pattern of activity and large-scale integration among regions rather than a specific brain region might be a critical condition for the neural correlate of consciousness.

A significant contribution to this discussion about the mechanisms of large-scale integration was reported by Crick and Koch (1992, 2003) in their studies on the visual consciousness. They suggested that the problem of binding might not be resolved only as a simple consequence of synchronization among large groups of neurons. As a basis for this opinion, they emphasize the binding problem of diffuse information representing a seen object by groups of synchronized, excited neurons that are located in different parts of the brain. This problem has emerged in connection with the finding that features of an object such as color, shape, texture, size, brightness, etc., produce activity in separate areas of the visual cortex (Crick and Koch 1992; Felleman and Van Essen 1991; Singer 1993, 2001). However, it is not known how spatial convergence is provided for the synthesis of the processed information that emerges; for example, there are only a few neural connections between specific visual areas that correlate with color and motion (Bartels and Zeki 2006; LaRock 2006; Zeki 1994, 2003).

Key evidence for the binding problem was reported in studies on the primate extrastriate visual cortex. These studies have shown that different neurons within the visual system participate in the processing of different features of a seen object (Desimone et al. 1985; Schein and Desimone 1990; Ghose and Tso 1997). Desimone et al. (1985) found that neurons in the visual area (V4) and the inferior temporal cortex (IT) are sensitive to many kinds of information relevant to object recognition. They also found that the special contribution of V4 neurons to visual processing may lie in the specific spatial and spectral interactions, and that many different stimulus qualities are handled in a parallel mode of processing. Ghose and Tso reported similar results (1997), and found that V4 contains modular assemblies of cells related to particular aspects of processed, object-based representations. There is also evidence in the research on moving objects that other neurons in the middle temporal area and the medial superior temporal area encode various aspects related to the motion of the stimulus (Treue and Andersen 1996; O'Keefe and Movshon 1998).

In the context of these findings, the hypothetical center for brain information convergence that enables perceptual consciousness and conscious experience was termed "Cartesian theater" (Crick and Koch 1992; Dennett 1991). Recent neuroscience, however, has not located a distinct place in which diffuse information in the brain comes together. Additionally, there is evidence that neocortical processing is distributed during all sensory and motor functions (Singer 1993, 2001).

The predominant view in the neuroscience of consciousness is that neuronal synchronization is a phenomenon that is necessary for the large-scale integration of diffuse neuronal activities. There is increasing experimental evidence that coherent neuronal assemblies in the brain are functionally linked by phase synchronization among simultaneously recorded EEG (electroencephalogram) signals, and that this time-dependent synchrony among various discrete neuronal assemblies denotes a neural substrate for mental representations, such as perception, cognitive functions, and memory (Varela et al. 2001; Lachaux et al. 1999). These functions are related to diffuse macroscopic patterns of neuronal activity, which involve multiple neuronal subsystems bound into a coherent whole (Braitenberg 1978; Van Putten and Stam 2001). According to recent data, a mechanism that enables binding of diffuse macroscopic patterns of neuronal activity, represented by neural assemblies, into a coherent whole, has still not been found and constitutes a fundamental problem in neuroscience (i.e., the binding problem, see above (Woolf and Hameroff 2001; Lee et al. 2003; Arp 2005; Fidelman 2005; Velik 2010). The theory of feature binding originates in distributed coding and hypothesizes that neurons involved in the processing of a single object will tend to synchronize firing, while simultaneously desynchronizing firing from the remaining neurons not involved in the processing of the object (von der Malsburg and Schneider 1986). An essential feature of neuronal assembly coding is that individual neurons or subsystems can participate at different times in an almost unlimited number of different assemblies (Sannita 2000; Varela et al. 2001). The same neurons can participate in different perceptual events and different combinations of these neurons can represent different perceptual objects. Synchronization of these different perceptual objects is related to the integration of perceptions into a coherent whole (Singer and Gray 1995). A candidate mechanism for the integration or binding of diffuse brain activities is the "gamma activity" – high-frequency oscillations of 40 Hz, but often varying from 30 to 90 Hz. This activity occurs synchronously across brain regions and underlies the integration of diverse brain activities (Singer and Gray 1995). Although most of the research on feature binding has focused on synchronous gamma activity, there is evidence that synchronous activities in other frequency bands might also participate in the functional integration of diffuse neural activities into the coherent whole (Bressler et al. 1993; Lee et al. 2003).

Following these findings Dennett (1991) proposed "a multiple drafts" theory of consciousness model that does not define consciousness as a unitary process, but rather as diffuse. Instead of a single central place, i.e., "Cartesian theater," there are various events of content fixation that occur in various places at various times in the brain (Dennett 1991, p. 365). The evidence for this view of consciousness presents a whole series of experimental results in cognitive neuroscience and psychology (LaRock 2006; Varela and Thompson 2003; von der Malsburg 1996, 1999; van der Velde and de Kamps 2006; Zeki 2003).

1.2 Anterior Cingulate Cortex and Mechanisms of Brain Integration

The majority of recent studies on the neural correlate of consciousness have focused on EEG analysis and have observed functionally relevant periods of synchronization, mainly in the gamma frequency band in various species and brain structures during attention, perception, motor, and memory tasks (Singer 2001; Lee et al. 2003; Jensen et al. 2007). Together, as mentioned above, these findings suggest that gamma activity (about 40 Hz) might be a candidate mechanism for the integration or binding of diffuse brain activities. The suggested mechanism of the gamma waves is that the wave, which originates in the thalamus, repeatedly oscillates 40 times per second in the brain back and forth, which enables different neuronal circuits to enter into synchrony with the perceptual information that is processed in the thalamus. This integrative process in turn enables these simultaneously active neuronal clusters to oscillate together during transient periods of synchronized firing and this coherent whole enables various memories and associations involved in the process to be connected and a coherent process of perception, cognitive processing, memory, and consciousness to be generated (Buzsaki 2006).

A crucial result of the studies mentioned above is the direct link between visual perception and gamma synchrony in the cat visual cortex, as reported by Eckhorn et al. (1988). Following this finding, the functional significance of synchronous gamma activity in selective attention, perceptual processing and recognition was repeatedly demonstrated in animal and human studies (Fries 2005; Meador et al. 2005; Rodriguez et al. 2004; Jensen et al. 2007). Together, these data strongly suggest that complex cognitive functions might be organized on a global level, which enables primitive functions organized in localized brain regions to be integrated

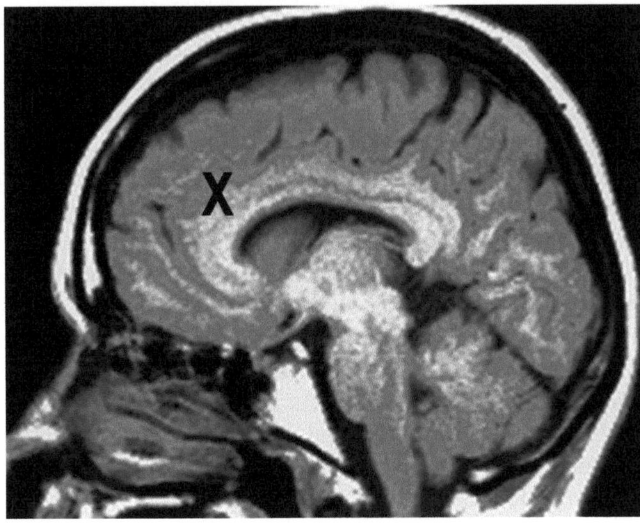

Fig. 1.2 Anterior cingulate cortex

(Bressler 2001). In this context, the current predominant hypothesis relies on the assumption of the global mode of functioning, which is based on large-scale information processing requiring mechanisms of functional integration of multiple disparate neural assemblies (Varela et al. 2001; Fries 2005; Jensen et al. 2007).

An important question for further research into the integration of diffuse brain activities is whether the essential integrative role can be attributed to a specific structure in the brain or whether this ability is inherent to the cognitive network as a whole. Recent concepts concerning the integrative role of the anterior cingulate cortex (ACC) seem to support the possibility that there could be some structural localization of integrative functions. In principle, this concept is supported by recent findings of anatomical connectivity of the ACC and its structural particularities, especially the presence of spindle-shaped neurons (Allman et al. 2001; Paus 2001; Posner et al. 2007; Kukleta et al. 2010). The spindle neurons represent a novel evolutionary specialization of the neural circuitry of the ACC (Allman et al. 2001). The circuitry containing the spindle cells may possess widespread connections with other parts of the brain and may also serve to coordinate the activity of these diverse parts to achieve self-control and the capacity to focus on difficult problems, such as emotional self-control and focused problem-solving, error recognition, and various adaptive responses to changing conditions (Allman et al. 2001; Posner et al. 2007). Current experimental evidence indicates that the role of the ACC in behavioral control comprises three main issues: motor control, cognition, and the motivation states of the organism. For example, there is evidence of increased functioning in the ACC in individuals with greater social insight and maturity. Lane and colleagues found that the activity of the ACC was greater in subjects who had higher levels of social awareness, based on objectively scored tests (Fig. 1.2) (Lane et al. 1998).

Recent findings on ACC specialization suggest that the functional involvement of the ACC in motor control, cognition, and motivation states of the organism might distinguish the ACC from other fronto-cortical regions and that this overlap might provide the ACC with the potential to translate intentions into actions (Paus 2001). A similar conceptual approach was developed by Rueda and Posner (Rueda et al. 2004; Posner et al. 2007), who based their proposal regarding the ACC role on patterns of ACC activation in studies in which withholding a dominant response was required to perform a subdominant response. They proposed that the major contribution of the ACC to brain functions relies on the ability to regulate information influx through orienting, in order to avoid conflicting responses in behavior (Rueda et al. 2004). In this context, there is also evidence that the ACC is involved in cognitive functions related to the processing and monitoring of conflicting information, which is related to higher ACC activation (Bunge et al. 2001; Paus 2001; Bob 2008). In addition, the specific integrative role of the ACC suggests a conceptual proposal regarding the central autonomic network, which includes structures that connect the brain structures with the autonomic nervous system. This network mainly includes the ACC, the insula, and the medial temporal lobe structures, such as the amygdala and hippocampus, which integrate emotional and cognitive information and exert a modulatory effect on the lower brain centers controlling the autonomic nervous system and modulating autonomic responses (Benarroch 1993; Bob et al. 2009a).

This specific integrative role of the ACC also seems to be evident from data suggesting that functional interactions manifesting at a level of EEG synchrony in the beta-2 frequency band might specifically distinguish the ACC from other fronto-temporal regions. This preliminary evidence indicates that EEG synchrony in the beta-2 frequency band (25–35 Hz), which is very close to gamma, distinguishes levels of synchrony between the ACC and various fronto-temporal areas from levels of synchrony among fronto-temporal regions (Kukleta et al. 2010). The crucial interpretation of this finding is that the higher level of synchrony of the pairs connecting the ACC electrode with a fronto-temporal electrode in comparison to pairs of fronto-temporal electrodes suggests that the ACC might play a particular role in large-scale communication, which could reflect its unique integrative functions in cognitive processing. This interpretation could have key consequences for understanding the neural correlate of consciousness, which, although it is spatially distributed and related to large-scale integration, may have its "extraordinary places" with a specific integrative role that enables conscious integration and cognitive functions.

1.3 Molecular Mechanisms of Brain Integration

1.3.1 Suprachiasmatic Nuclei and Melatonin

Neurons of suprachiasmatic nuclei (SCN) of the hypothalamus play a specific and major role in the processes of temporal integration and binding related to consciousness, cognitive processes, and memory function. The temporal patterns of rhythmicity in the SCN are generated by gene expression of individual SCN neurons.

These individual neural oscillators are organized into coherent activity of the biological clock and facilitate temporal synchronization, which produces differentially timed waves specifically targeting the pineal gland and other structures, and controlling neuroendocrine rhythms (Kalsbeek et al. 2006; Indic et al. 2007; Hamada et al. 2004).

Among endocrine output signals related to circadian information processing, melatonin plays an important role as an endogenous synchronizer that is able to stabilize and reinforce circadian rhythms and maintain their mutual phase relationship. This integrative process occurs at different levels of the circadian network via gene expression in some brain regions and peripheral structures, and enables integration of circadian, hormonal, and metabolic information, creating temporal order of bodily and mental experience (Rutter et al. 2002; Pevet et al. 2006; Saper et al. 2005). This specific temporal order is reflected in the associative process necessary for cognition, behavior, and all processes of memory consolidation that must preserve information in the temporal causal order and synchrony or sequentiality of the internal cognitive maps. In this context recent findings suggest that melatonin could be a potential regulator in the processes that contribute to memory formation, long-term potentiation (LTP), and synaptic plasticity in the hippocampus and other brain regions (Lynch 2004; Baydas et al. 2005; Larson et al. 2006; Chaudhury et al. 2005; Ozcan et al. 2006; Gorfine and Zisapel 2007).

The basic mechanism of melatonin action consists of a possible interaction with both excitatory and inhibitory neurotransmitter systems (Larson et al. 2006; Saenz et al. 2004; Skaper et al. 1998). One mechanism that is likely to underlie the effects of melatonin on synaptic plasticity is modulation of the intrinsic excitability of the hippocampal neurons. Hyperpolarization induced by melatonin could reduce long-term potentiation (LTP) by inhibiting NMDA receptor activation during high-frequency stimulation (Wang et al. 2005). Melatonin application decreases membrane excitability in other regions of the nervous system, in part via an enhancement of potassium currents (Wang et al. 2005). Melatonin may also decrease the potential generation of spontaneous action in the SCN (Shibata et al. 1989; Stehle et al. 1989; Mason and Rusak 1990) through an increase in potassium conductance and a decrease in a hyperpolarization-activated current (Jiang et al. 1995; van den Top et al. 2001).

Melatonin may also inhibit LTP induction through regulation of signaling pathways downstream of the membrane and NMDA receptor activation. Outside of the hippocampus, melatonin may influence rhythms in gene expression and second messenger systems (von Gall et al. 2002; Gerdin et al. 2004; Wang et al. 2005). Electrophysiological studies have also demonstrated that melatonin can regulate the electrical activity of hippocampal neurons (Zeise and Semm 1985; Musshoff et al. 2002) and alter synaptic transmission between hippocampal neurons (Wan et al. 1999; Hogan et al. 2001; El-Sherif et al. 2003). These findings indicate that melatonin can regulate learning and memory through its influence on synaptic connections within the hippocampus undergoing activity-dependent changes in synaptic strength, including enhancements in the strength of the excitatory synaptic transmission that regulates LTP. In this context there is evidence that stress disrupts normal activity and memory consolidation in the hippocampus and prefrontal cortex (Diamond and Rose 1994; Ruel and de Kloet 1985; Payne et al. 2006). This process leads to memories that are stored without a contextual or spatiotemporal frame and produces memories that are often fragmentary and temporally and spatially disorganized, mainly because they originate from entirely unrelated events (Nadel and Jacobs 1998). Disturbed temporal memory related to stress conditions is evident from studies focused on episodic and autobiographical memories and shows temporal disorganization, fragmentation, and incompleteness, but not necessarily an absence of emotion or dissociative amnesia (Kenardy et al. 2007; Payne et al. 2006; Brewin 2007).

According to recent experimental findings cognitive processing associated with stress events is related to melatonin alterations in animals and in humans. For example, repeated maternal separation and deprivation caused low blood melatonin levels and a significantly negative correlation between blood melatonin levels and spatial memory performance in both male and female adolescent rats, which suggests an association between melatonin production and neurodevelopment (Uysal et al. 2005). Further studies also found an interaction between stress and the pineal gland (Vollrath and Welker 1988; Simonneaux and Ribelayga 2003), and some electron microscopy studies have found that immobilization stress induces pinealocyte degeneration (Milin et al. 1996). Psychosocial stress may also induce a robust increase in melatonin metabolite 6-sulfatoxymelatonin in subordinate animals (Fuchs and Schumacher 1990). In humans, stress may cause sleep disturbances, such as insomnia, and a reduced nocturnal peak of pineal melatonin secretion, which is often present in depressed patients (Jindal and Thase 2004; Brown et al. 1985; Frazer et al. 1986; Pacchierotti et al. 2001). These studies suggest that the pineal gland might be significantly affected by stress, which is consistent with findings reporting that the pineal gland expresses a high density of the glucocorticoid receptor (Warembourg 1975; Sarrieau et al. 1988; Meyer et al. 1998). In addition, melatonin receptors are also present in regions that participate in the stress response, such as the hippocampus or the adrenal gland (Musshoff et al. 2002; Torres-Farfan et al. 2003).

Together, these findings suggest that melatonin is likely significantly associated with the regulation of memory, cognition, and emotional processes (Laudon et al. 1989; Boatright et al. 1994; Hemby et al. 2003). These findings emphasize the specific role of melatonin in the mechanisms of cognition, memory, and stress, and are consistent with reported studies that indicate melatonin alterations in psychopathology, mainly in patients with depression, schizophrenia, anxiety disorders, eating disorders, and other mental disorders (Pacchierotti et al. 2001; Bob and Fedor-Freybergh 2008).

1.3.2 Brain-Derived Neurotrophic Factor

Brain-derived neurotrophic factor (BDNF) is a polypeptide growth factor that influences differentiation and survival of neurons in the nervous system and is important in regulating synaptic plasticity and connectivity in the CNS, with implications for mechanisms of consciousness, cognition, memory storage, and mood control (Bath and Lee 2006; Bramham and Messaoudi 2005). BDNF acts on certain neurons of the central nervous system and on the peripheral nervous system, helping existing neurons to survive, and encouraging the growth and differentiation of new neurons and synapses (Acheson et al. 1995). In the brain, it is mainly active in the areas of the hippocampus, cortex, and basal forebrain, and plays an important role in learning, memory, and higher thinking (Yamada and Nabeshima 2003). BDNF is an activity-dependent modulator of excitatory transmission and synaptic plasticity and, together with its receptor, tyrosine kinase TrkB, is predominantly located on glutamate synapses (Soule et al. 2006; Bramham and Messaoudi 2005). Recent evidence indicates that endogenous BDNF-TrkB signaling in synaptic consolidation by LTP requires new gene expression and protein synthesis, which enables the immediate early gene, Arc (activity-regulated cytoskeleton-associated protein) (Bath and Lee 2006; Bramham and Messaoudi 2005; Soule et al. 2006). An important factor in this new gene expression is the transcription factor CREB, which is required for hippocampus-dependent, long-term memory formation (Nadel 1994; Mizuno and Giese 2005). The CREB is activated by signaling pathways that include Ca(2+)/ calmodulin kinases (CaMKs), protein kinase A (PKA), and the mitogen-activated protein/extracellular signal-regulated kinases (MAPK or ERKs) (Mizuno and Giese 2005; Rattiner et al. 2005). Recent molecular genetic and behavioral studies also demonstrate that spatial and contextual types of hippocampus-dependent formation of long-term memory require different signaling molecules, implicating distinct types of hippocampus-dependent long-term memory that have different underlying molecular mechanisms (Mizuno and Giese 2005). As a part of these signaling pathways a basic mechanism of BDNF is the possible modulation of both excitatory and inhibitory neurotransmitter systems (Savitz et al. 2006). According to several studies BDNF also influences the functions of the serotonergic and dopaminergic systems (Savitz et al. 2006; Narita et al. 2003; Mossner et al. 2000).

The relationship between BDNF and cognition is also mediated by the influence of stress (Savitz et al. 2006). Chronic stress, in particular, influences excessive release of glucocorticoids from the adrenal gland that cause cell death or atrophy of vulnerable neurons through the action of cortisol and the inhibitory influence on BDNF synthesis in the hippocampus (Savitz et al. 2006). Recent findings have suggested that BDNF might play an important role in the stress response and the related modification of synaptic plasticity, transmission, and memory formation, especially in the hippocampus and neocortex. It also plays a specific role in depression, schizophrenia, epilepsy, neurodegenerative disorders, and pain sensitization (Binder and Scharfman 2004; Thomas and Davies 2005).

References

Acheson A, Conover JC, Fandl JP, DeChiara TM, Russell M, Thadani A, Slinto SP, Yancopoulos GD, Lindsay RM. A BDNF autocrine loop in adult sensory neurons prevents cell death. Nature. 1995;374:450–3.

Allman JM, Hakeem A, Erwin JM, Nimchinsky E, Hof P. The anterior cingulate cortex. The evolution of an interface between emotion and cognition. Ann N Y Acad Sci. 2001;935:107–17.

Arp R. Selectivity, integration, and the psycho-neuro-biological continuum. J Mind Behav. 2005;26:35–64.
Barrera-Mera B, Barrera-Calva E. The Cartesian clock metaphor for pineal gland operation pervades the origin of modern chronobiology. Neurosci Biobehav Rev. 1998;23:1–4.
Bartels A, Zeki S. The temporal order of binding visual attributes. Vision Res. 2006;46:2280–6.
Bath KG, Lee FS. Variant BDNF (Val66Met) impact on brain structure and function. Cogn Affect Behav Neurosci. 2006;6:79–85.
Baydas G, Ozveren F, Akdemir I, Tuzcu M, Yasar A. Learning and memory deficits in rats induced by chronic thinner exposure are reversed by melatonin. J Pineal Res. 2005;39:50–6.
Benarroch EE. The central autonomic network — functional organization, dysfunction and perspective. Mayo Clin Proc. 1993;68:988–1001.
Binder DK, Scharfman HE. Brain-derived neurotrophic factor. Growth Factors. 2004;22:123–31.
Boatright JH, Rubim NM, Iuvone PM. Regulation of endogenous dopamine release in amphibian retina by melatonin: the role of GABA. Vis Neurosci. 1994;11:1013–8.
Bob P. Pain, dissociation and subliminal self-representations. Conscious Cogn. 2008;17:355–69.
Bob P. Quantum science and the nature of mind. J Mind Behav. 2009;30:1–14.
Bob P, Fedor-Freybergh P. Melatonin, consciousness and traumatic stress. J Pineal Res. 2008;44:341–7.
Bob P, Susta M, Gregusova A, Jasova D. Dissociation, cognitive conflict and nonlinear patterns of heart rate dynamics in patients with unipolar depression. Prog Neuropsychopharmacol Biol Psychiatry. 2009;33:141–5.
Bogen JE. On the neurophysiology of consciousness: part I- an overview. Conscious Cogn. 1995;4:52–62.
Braitenberg V. Cell assemblies in the cerebral cortex. In: Heim R, Palm G, editors. Theoretical approaches to complex systems. Lecture notes in biomathematics. Berlin: Springer; 1978. p. 171–88.
Bramham CR, Messaoudi E. BDNF function in adult synaptic plasticity: the synaptic consolidation hypothesis. Prog Neurobiol. 2005;76:99–125.
Bressler, SL, Kelso JAS. Cortical coordination dynamics and cognition. Trends Cogn Sci. 2001;5:26–6.
Bressler SL, Coppola R, Nakamura R. Episodic multiregional cortical coherence at multiple frequencies during visual task performance. Nature. 1993;366:153–6.
Brewin CR. Autobiographical memory for trauma: update on four controversies. Memory. 2007;15:227–48.
Brown R, Kocsis JH, Caroff S. Differences in nocturnal melatonin secretion between melancholic depressed patients and control subjects. Am J Psych. 1985;142:811–16.
Bunge SA, Ochsner KN, Reskond JE, Dover GH, Gabrieli JDE. Prefrontal regions involved in keeping information in and out of mind. Brain. 2001;124:2074–86.
Buzsaki G. Rhythms of the brain. Oxford: Oxford University Press; 2006.
Chaudhury D, Wang LM, Colwell CS. Circadian regulation of hippocampal long-term potentiation. J Biol Rhythms. 2005;20:225–36.
Cotterill RMJ. On the unity of conscious experience. J Conscious Stud. 1995;2:290–12.
Crick F, Koch C. The problem of consciousness. Sci Am. 1992;267(3):153–9.
Crick F, Koch C. A framework for consciousness. Nat Neurosci. 2003;6:119–26.
Dennett D. Consciousness explained. Boston: Little, Brown; 1991.
Desimone R, Schein SJ, Moran J, Ungerleider LG. Contour, color and shape analysis beyond the striate cortex. Vision Res. 1985;25:441–52.
Diamond DM, Rose GM. Stress impairs LTP and hippocampal-dependent memory. Ann N Y Acad Sci. 1994;746:411–4.
Eckhorn R, Bauer R, Jordan W, Brosch M, Kruse W, Munk M, Reitboeck HJ. Coherent oscillations: a mechanism of feature linking in the visual cortex? Biol Cybern. 1988;60:121–30.
El-Sherif Y, Tesoriero J, Hogan MV, Wieraszko A. Melatonin regulates neuronal plasticity in the hippocampus. J Neurosci Res. 2003;72:454–60.
Felleman D, Van Essen D. Distributed hierarchical processing in the primate visual cortex. Cereb Cortex. 1991;1:1–47.

Fidelman U. Visual search and quantum mechanics: a neuropsychological basis of Kant's creative imagination. J Mind Behav. 2005;26:23–33.
Finger S. Origins of neuroscience: a history of explorations into brain function. New York: Oxford University Press; 1994.
Frazer A, Brown R, Kocsis J, Caroff S, Amsterdam J, Winokur A, Sweeney J, Stokes P. Patterns of melatonin rhythms in depression. J Neural Transm Suppl. 1986;21:269–90.
Fries P. A mechanism for cognitive dynamics: neuronal communication through neuronal coherence. Trends Cogn Neurosci. 2005;9:474–80.
Fuchs E, Schumacher M. Psychosocial stress affects pineal function in the tree shrew (Tupaia belangeri). Physiol Behav. 1990;47:713–7.
Gerdin MJ, Masana MI, Rivera-Bermudez MA, Hudson RL, Earnest DJ, Gillette MU, Dubocovich ML. Melatonin desensitizes endogenous MT2 melatonin receptors in the rat suprachiasmatic nucleus: relevance for defining the periods of sensitivity of the mammalian circadian clock to melatonin. FASEB J. 2004;18:1646–56.
Ghose GM, Tso DY. Form processing modules in primate area V4. J Neurophysiol. 1997;77:2191–6.
Gorfine T, Zisapel N. Melatonin and the human hippocampus, a time dependent interplay. J Pineal Res. 2007;43:80–6.
Gray JA. The contents of consciousness: a neuropsychological conjecture. Behav Brain Sci. 1995;18:659–22.
Hamada T, Antle MC, Silver R. Temporal and spatial expression patterns of canonical clock genes and clock-controlled genes in the suprachiasmatic nucleus. Eur J Neurosci. 2004;19:1741–8.
Hemby SE, Trojanowski JQ, Ginsberg SD. Neuron-specific age-related decreases in dopamine receptor subtype mRNAs. J Comp Neurol. 2003;456:176–83.
Hogan MV, El-Sherif Y, Wieraszko A. The modulation of neuronal activity by melatonin: in vitro studies on mouse hippocampal slices. J Pineal Res. 2001;30:87–96.
Indic P, Schwartz WJ, Herzog ED, Foley NC, Antle MC. Modeling the behavior of coupled cellular circadian oscillators in the suprachiasmatic nucleus. J Biol Rhythms. 2007;22:211–9.
Jensen O, Kaiser J, Lachaux JP. Human gamma-frequency oscillations associated with attention and memory. Trends Neurosci. 2007;30:317–24.
Jiang ZG, Nelson CS, Allen CN. Melatonin activates an outward current and inhibits Ih in rat suprachiasmatic nucleus neurons. Brain Res. 1995;687:125–32.
Jindal RD, Thase ME. Treatment of insomnia associated with clinical depression. Sleep Med Rev. 2004;8:19–30.
John ER. The neurophysics of consciousness. Brain Res Rev. 2002;39:1–28.
Kalsbeek A, Palm IF, La Fleur SE, Scheer FA, Perreau-Lenz S, Ruiter M, Kreier F, Cailotto C, Buijs RM. SCN outputs and the hypothalamic balance of life. J Biol Rhythms. 2006;21:458–69.
Kandel E, Schwarz J, Jessel T. Principles of neural science. New York: McGraw–Hill; 2000.
Kenardy J, Smith A, Spence SH, Lilley PR, Newcombe P, Dob R, Robinson S. Dissociation in children's trauma narratives: an exploratory investigation. J Anxiety Disord. 2007;21:456–66.
Kukleta M, Bob P, Brázdil M, Roman R, Rektor I. The level of frontal-temporal beta-2 band EEG synchronization distinguishes anterior cingulate cortex from other frontal regions. Conscious Cogn. 2010;19:879–86.
Lachaux JP, Rodriguez E, Martinerie J, Varela FJ. Measuring phase synchrony in brain signals. Hum Brain Mapp. 1999;8:194–208.
Lane RD, Reiman EM, Axelrod B, Yun LS, Holmes A, Schwarz GE. Neural correlates of levels of emotional awareness. Evidence of an interaction between emotion and attention in the anterior cingulate cortex. J Cogn Neurosci. 1998;10:525–35.
LaRock E. Why neural synchrony fails to explain the unity of visual consciousness. Behav Philos. 2006;34:39–58.
Larson J, Jessen RE, Uz T, Arslan AD, Kurtuncu M, Imbesi M, Manev H. Impaired hippocampal long-term potentiation in melatonin MT2 receptor-deficient mice. Neurosci Lett. 2006;393:23–6.

Laudon M, Hyde JF, Ben-Jonathan N. Ontogeny of prolactin releasing and inhibiting activities in the posterior pituitary of male rats. Neuroendocrinology. 1989;50:644–49.

Lee KH, Williams LM, Breakspear M, Gordon E. Synchronous gamma activity: a review and contribution to an integrative neuroscience model of schizophrenia. Brain Res Rev. 2003;41:57–78.

Lynch MA. Long-term potentiation and memory. Physiol Rev. 2004;84:87–136.

Mason R, Rusak B. Neurophysiological responses to melatonin in the SCN of short-day sensitive and refractory hamsters. Brain Res. 1990;533:15–19.

Meador KJ, Ray PG, Echauz JR, Loring DW, Vachtsevanos GJ. Gamma coherence and conscious perception. Neurology. 2005;59:1847–54.

Meyer U, Kruhoffer M, Flugge G, Fuchs E. Cloning of glucocorticoid receptor and mineralocorticoid receptor cDNA and gene expression in the central nervous system of the tree shrew (Tupaia belangeri). Brain Res Mol Brain Res. 1998;55:243–53.

Milin J, Demajo M, Todorovic V. Rat pinealocyte reactive response to a long-term stress inducement. Neuroscience. 1996;73:845–54.

Mizuno K, Giese KP. Hippocampus-dependent memory formation: do memory type-specific mechanisms exist? J Pharmacol Sci. 2005;98:191–7.

Mossner R, Daniel S, Albert D, Heils A, Okladnova O, Schmitt A, Lesch KP. Serotonin transporter function is modulated by brain derived neurotrophic factor (BDNF) but not nerve growth factor (NGF). Neurochem Int. 2000;36:197–202.

Musshoff U, Riewenherm D, Berger E, Fauteck JD, Speckmann EJ. Melatonin receptors in rat hippocampus: molecular and functional investigations. Hippocampus. 2002;12:165–73.

Nadel L. Hippocampus, space, and relations. Behav Brain Sci. 1994;17:490–91.

Nadel L, Jacobs WJ. Traumatic memory is special. Curr Dir Psychol Sci. 1998;7:154–57.

Narita M, Aoki K, Takagi M, Yajima Y, Suzuki T. Implication of brain derived neurotrophic factor in the release of dopamine and dopamine-related behaviors induced by amphetamine. Neuroscience. 2003;119:767–75.

O'Keefe LP, Movshon JA. Processing of first- and second-order motion signals by neurons in area MT of the macaque monkey. Vis Neurosci. 1998;15:305–17.

Ozcan M, Yilmaz B, Carpenter DO. Effects of melatonin on synaptic transmission and long-term potentiation in two areas of mouse hippocampus. Brain Res. 2006;1111:90–4.

Pacchierotti C, Iapichino S, Bossini L, Pieraccini F, Castrogiovanni P. Melatonin in psychiatric disorders: a review on the melatonin involvement in psychiatry. Front Neuroendocrinol. 2001;22:18–32.

Paus T. Primate anterior cingulate cortex: where motor control, drive and cognition interface. Nat Rev Neurosci. 2001;2:417–24.

Payne JD, Jackson ED, Ryan L, Hoscheidt S, Jacobs JW, Nadel L. The impact of stress on neutral and emotional aspects of episodic memory. Memory. 2006;14:1–16.

Pevet P, Agez L, Bothorel B, Saboureau M, Gauer F, Laurent V, Masson-Pevet M. Melatonin in the multi-oscillatory mammalian circadian world. Chronobiol Int. 2006;23:39–51.

Posner MI, Rothbart MK, Sheese BE, Tang Y. The anterior cingulate gyrus and the mechanism of self-regulation. Cogn Affect Behav Neurosci. 2007;7:391–95.

Rattiner LM, Davis M, Ressler KJ. Brain-derived neurotrophic factor in amygdala-dependent learning. Neuroscientist. 2005;11:323–33.

Rodriguez R, Kallenbach U, Singer W, Munk MH. Short- and long-term effects of cholinergic modulation on gamma oscillations and response synchronization in the visual cortex. J Neurosci. 2004;24:10369–78.

Rueda MR, Posner MI, Rothbart MK. Attentional control and self-regulation. In: Baumeister RF, Vohs KD, editors. Handbook of self-regulation: research, theory, and applications. New York: Guilford; 2004. p. 283–300.

Ruel JMHM, de Kloet ER. Two receptor systems for corticosterone in rat brain: microdistribution and differential occupation. Endocrinology. 1985;117:2505–12.

Rutter J, Reick M, McKnight SL. Metabolism and the control of circadian rhythms. Annu Rev Biochem. 2002;71:307–31.

Saenz DA, Goldin AP, Minces L, Chianelli M, Sarmiento MI, Rosenstein RE. Effect of melatonin on the retinal glutamate/glutamine cycle in the golden hamster retina. FASEB J. 2004;18: 1912–13.

Sannita W. Stimulus-specific oscillatory responses of the brain: a time/frequency related coding process. Clin Neurophysiol. 2000;11:565–83.

Saper CB, Lu J, Chou TC, Gooley J. The hypothalamic integrator for circadian rhythms. Trends Neurosci. 2005;28:152–7.

Sarrieau A, Dussaillant M, Moguilewsky M. Autoradiographic localization of glucocorticosteroid binding sites in rat brain after in vivo injection of [3H]RU 28362. Neurosci Lett. 1988;92:14–20.

Savitz J, Solms M, Ramesar R. The molecular genetics of cognition: dopamine, COMT and BDNF. Genes Brain Behav. 2006;5:311–28.

Schein SJ, Desimone R. Spectral properties of V4 neurons in the macaque. J Neurosci. 1990;10:3369–89.

Shibata S, Cassone VM, Moore RY. Effects of melatonin on neuronal activity in the rat suprachiasmatic nucleus in vitro. Neurosci Lett. 1989;97:140–4.

Simonneaux V, Ribelayga C. Generation of the melatonin endocrine message in mammals: a review of the complex regulation of melatonin synthesis by norepinephrine peptides, and other pineal transmitters. Pharmacol Rev. 2003;55:325–95.

Singer W. Synchronization of cortical activity and its putative role in information procesing and learning. Annu Rev Physiol. 1993;55:349–74.

Singer W. Consciousness and the binding problem. Ann N Y Acad Sci. 2001;929:123–46.

Singer W, Gray CM. Visual feature integration and the temporal correlation hypothesis. Annu Rev Neurosci. 1995;18:555–86.

Skaper SD, Ancona B, Facci L, Franceschini D, Giusti P. Melatonin prevents the delayed death of hippocampal neurons induced by enhanced excitatory neurotransmission and the nitridergic pathway. FASEB J. 1998;12:725–31.

Smith CUM. Descartes' pineal neuropsychology. Brain Cogn. 1998;36:57–72.

Soule J, Messaoudi E, Bramham CR. Brain-derived neurotrophic factor and control of synaptic consolidation in the adult brain. Biochem Soc Trans. 2006;34:600–4.

Stehle J, Vanecek J, Vollrath L. Effects of melatonin on spontaneous electrical activity of neurons in rat suprachiasmatic nuclei: an in vitro iontophoretic study. J Neural Transm. 1989;78:173–77.

Thomas K, Davies A. Neurotrophins: a ticket to ride for BDNF. Curr Biol. 2005;15:262–4.

Torres-Farfan C, Richter HG, Rojas-García P, Vergara M, Forcelledo ML, Valladares LE, Torrealba F, Valenzuela GJ, Serón-Ferré M. mt1 melatonin receptor in the primate adrenal gland: inhibition of adrenocorticotropin-stimulated cortisol production by melatonin. J Clin Endocrinol Metab. 2003;88:450–58.

Treue S, Andersen RA. Neural responses to velocity gradients in macaque cortical area MT. Vis Neurosci. 1996;13:797–804.

Uysal N, Ozdemir D, Dayi A, Yalaz G, Baltaci AK, Bediz CS. Effects of maternal deprivation on melatonin production and cognition in adolescent male and female rats. Neuro Endocrinol Lett. 2005;26:555–60.

van de Grind W. Physical, neural, and mental timing. Conscious Cogn. 2002;11:241–64.

van den Top M, Buijs RM, Ruijter JM, Delagrange P, Spanswick D, Hermes ML. Melatonin generates an outward potassium current in rat suprachiasmatic nucleus neurones in vitro independent of their circadian rhythm. Neuroscience. 2001;107:99–108.

van der Velde F, de Kamps M. Neural blackboard architectures of combinatorial structures in cognition. Behav Brain Sci. 2006;29:1–72.

Van Putten MJAM, Stam CJ. Is the EEG really "chaotic" in hypsarrhythmia. IEEE Eng Med Biol Mag. 2001;20:72–79.

Varela F, Thompson E. Neural synchrony and the unity of mind: a neurophenomenological perspective. In: Cleeremans A, editor. The unity of consciousness: binding, integration, and dissociation. Oxford: Oxford University Press; 2003. p. 266–87.

Varela FJ, Lachaux JP, Rodriguez E, Martinerie J. The brainweb: phase synchronization and large-scale integration. Nat Rev Neurosci. 2001;2:229–39.

Velik R. From single neuron-firing to consciousness- towards the true solution of the binding problem. Neurosci Biobehav Rev. 2010;34:993–1001.

von der Malsburg C. The binding problem of neural networks. In: Llinas R, Churchland PS, editors. The mind-brain continuum: sensory processes. Cambridge, Massachusetts: MIT Press; 1996. p. 131–46.

von der Malsburg C. The what and why of binding: the modeler's perspective. Neuron. 1999;24: 95–104.

von der Malsburg C, Schneider W. A neural cocktail-party processor. Biol Cybern. 1986;54: 29–40.

von Gall C, Garabette ML, Kell CA, Frenzel S, Dehghani F, Schumm-Draeger PM, Weaver DR, Korf HW, Hastings MH, Stehle JH. Rhythmic gene expression in pituitary depends on heterologous sensitization by the neurohormone melatonin. Nat Neurosci. 2002;5:234–38.

Vollrath L, Welker HA. Day-to-day variation in pineal serotonin N-acetyltransferase activity in stressed and nonstressed male Sprague-Dawley rats. Life Sci. 1988;42:2223–29.

Wan Q, Man HY, Liu F, Braunton J, Niznik HB, Pang SF, Brown GM, Wang YT. Differential modulation of GABAA receptor function by Mel1a and Mel1b receptors. Nat Neurosci. 1999;2:401–03.

Wang LM, Suthana NA, Chaudhury D, Weaver DR, Colwell CS. Melatonin inhibits hippocampal long-term potentiation. Eur J Neurosci. 2005;22:2231–37.

Warembourg M. Radioautographic study of the rat brain and pituitary after injection of 3H dexamethasone. Cell Tissue Res. 1975;161:183–191.

Weiskrantz L. Consciousness and commentaries. In: Hameroff SR, Kaszriak A, Scott AC, editors. Toward a science of consciousness II—The second Tucson discussions and debates. Cambridge, MA: MIT Press; 1998. p. 371–77.

Woolf N, Hameroff S. A quantum approach to visual consciousness. Trends Cogn Neurosci. 2001;5:472–78.

Yamada K, Nabeshima T. Brain-derived neurotrophic factor/TrkB signaling in memory processes. J Pharmacol Sci. 2003;91:267–70.

Zeise ML, Semm P. Melatonin lowers excitability of guinea pig hippocampal neurons in vitro. J Comp Physiol. 1985;57:23–29.

Zeki S. A vision of the brain. Oxford: Blackwell Science; 1994.

Zeki S. The disunity of consciousness. Trends Cogn Neurosci. 2003;7:214–18.

Zeman A. Consciousness. Brain. 2001;124(Pt 7):1263–89.

Chapter 2
Binding Problem of Consciousness

This chapter describes several candidate mechanisms that might explain the binding of distributed macroscopic patterns of neuronal activities into a coherent whole. According to current findings, this problem is still unresolved and represents a fundamental problem in neuroscience related to brain coding and integration of distributed neural activities during processes related to perception, cognition, and memory (the "binding problem"). The theory of feature binding originated in the concept of distributed coding and states that neurons involved in the processing of a single object will tend to synchronize firing. This neural activity occurs synchronously across brain regions and likely underlies the integration of diverse brain activities. Together, these findings indicate that the solution to the binding problem may lie in the fundamental problem of consciousness in modern neuroscience. The predominant opinion is that consciousness emerges from a dynamical nucleus of persisting reverberation and interactions of neural groups. Other approaches to the binding problem include nonconventional hypotheses related to various physical theories, such as the complexity theory and the chaos theory.

According to a recent growing body of evidence, the neural correlate of consciousness is related to the processing of distributed information that is represented by integration through levels of neural synchronization among multiple brain regions, which is in turn related to large-scale integration, or "binding." A seminal contribution to discussions about mechanisms of large-scale integration was made by Crick and Koch (1992, 2003), who proposed that the problem of binding cannot simply be resolved as a simple consequence of synchronization among large groups of neurons. As a basis for this opinion they emphasize the binding problem of distributed information represented by different modalities (such as form, motion, color, smell, sound). Processing information related to the perceived object produces synchronous activities in separate areas of the brain, but there is no evidence of the spatial convergence in the brain that would represent the neural correlate of consciousness. The hypothetical center for information convergence was termed "Cartesian theater" (Crick and Koch 1992; Dennett 1991), but recent neuroscience has not located a distinct place in which information distributed in the brain comes together. Recent findings suggest that a candidate mechanism for the integration

P. Bob, *Brain, Mind and Consciousness: Advances in Neuroscience Research*,
DOI 10.1007/978-1-4614-0436-1_2, © Springer Science+Business Media, LLC 2011

or binding of distributed brain activities might be the "gamma activity" (most frequently high-frequency oscillations of about 40 Hz). Although there is growing evidence that EEG gamma waves enable different neuronal circuits to enter into synchrony with the perceptual information and oscillate together during transient periods of synchronized firing, there is no explanation of what mechanism is behind this synchronization and information convergence, integrating various percepts, memories, and associations. From this point of view the phenomenon of synchronization and functional integration presents evidence that information convergence is successfully achieved, but what is behind this process of synchronization remains a puzzle.

As already stated above, a solution to the binding problem may be found within the fundamental problem of consciousness. For example, Tononi and Edelman (2000) emphasize that consciousness is the re-entry of neural signals via changes in complexity and entropy in the central nervous system. Libet (1998) suggests that subjective experience might represent a field emerging from neural synchronization and coherence, and is not reducible to any physical process (see also John 2002). In accordance with Libet, Squires (1998) maintains that consciousness can be understood to be a primitive (irreducible) component of the world, and includes specific qualities of subjective experience (*qualia*) that cannot be reduced to any other physical quality (see also Duch 2005; John 2002). According to Freeman (1991, 2000, 2001), the image of the world that we have emerges as a consequence of creating order from nonlinear chaotic activity of large groups of neurons. These nonlinear chaotic processes represent a consequence of high system complexity, when the system involves a large number of complex interlinked and simultaneously active neural assemblies and runs in a desynchronized mode of parallel distribution that can lead to self-organization (Freeman 1991, 2000, 2001; Velazquez et al. 2003) and typical dynamical instabilities in mental phenomena (Atmanspacher and Fach 2005).

2.1 Consciousness and Complexity

Detailed studies on the mechanism of consciousness, binding mechanism, and complexity were reported by Tononi and Edelman (1998) and in later studies by Sporns et al. (2000, 2002). In contrast to conventional approaches to understanding consciousness, which are generally concerned with the contribution of specific brain areas or groups of neurons, Tononi and Edelman (1998) instead found out what kinds of neural processes represent the key properties of conscious experience. They applied measures of neural integration and complexity, and proposed the dynamic core hypothesis with regard to the properties of the neural substrate of consciousness. In agreement with the usual evidence, Tononi and Edelman (1998) postulated that conscious experience is integrated (i.e., each conscious scene is unified) but also highly differentiated by a huge number of differently experienced states of consciousness. They provided tools for the measurement of integration linked to functional clustering and differentiation representing neural complexity, which can be applied to actual neural processes.

Tononi and Edelman (1998) also proposed that consciousness is not a thing, but a process or changing stream on a time scale of fractions of seconds. In agreement with William James (1890), they emphasized that a fundamental aspect of consciousness is an integrated state, although at the same time there is evidence that distributed neural activity, particularly in the thalamocortical system, is essential for conscious experience (Edelman 1989; Picton and Stuss 1994; Newman 1995). Therefore, it is possible to suppose that interactions among neuronal assemblies in distributed brain areas might create a unified neural process corresponding to a multimodal conscious scene (Edelman 1989). In this context, Tononi and Edelman (1998) suggested that a key neural mechanism underlying conscious experience might be represented by interactions between posterior thalamocortical areas, which are involved in perceptual categorization, and anterior thalamocortical areas, which are related to memory processing and action planning. Furthermore, Tononi and Edelman (1998) suggested that such rapid interactions might be achieved through the process of re-entry, i.e., the ongoing, recursive, highly parallel signaling within and among brain areas. Using large-scale computer simulations, they showed that re-entry can achieve dynamical integration or "binding" of distributed and functionally specialized neuronal groups in a unified neural process without a single place – "Cartesian theater" (Tononi et al. 1992; Lumer et al. 1997; Tononi and Edelman 1998). In agreement with this current evidence changes in conscious experience driven by external stimuli, memories, mental images or dreams are also related to changes in the activity or deactivation of specific, widely distributed brain areas (Roland 1993; Frackowiak 1997; Tononi and Edelman 1998).

Some modeling studies suggest that a specific sign of effective re-entrant interactions might be short-term temporal correlations between the neuronal groups involved (Tononi and Edelman 1998). Other studies indicate that various kinds of cognitive tasks are related to short-term temporal correlations among distributed populations of neurons in the thalamocortical system (Bressler 1995; Tononi and Edelman 1998). In addition, reported magnetoencephalographic study of binocular rivalry indicates that awareness of a stimulus is related to increased coherence among distant brain regions (Tononi et al. 1998a). This condition of fast, strong, and distributed neural interactions related to binding may explain why feeble, degraded, or short-lasting stimuli are often not consciously perceived, even though they may produce a behavioral response, such as perception, without awareness (Marcel 1983; Merikle et al. 2001).

Tononi and Edelman (1998) applied methods of functional clustering and found that a subset of distributed elements within a system gives rise to a single, integrated process in cases where these elements interact significantly more strongly among themselves than with the other parts of the system. This interaction means that they form a functional cluster that can be measured by mutual information as a level of integration (Tononi et al. 1998b). When the level of integration calculated among all neurons within the subsystem is higher than the level of integration that the same neurons of this subsystem have with neurons outside of the subsystem, then the subsystem presents the functional cluster (Papoulis 1991; Tononi and Edelman 1998). For example, it is possible to compare levels of synchronous firing among cortical regions and between the cortex and thalamus (Tononi et al. 1992; Lumer et al. 1997; Tononi and Edelman 1998). Functional clustering also enables system complexity to be defined as a number of its parts (i.e., number of clusters) that have a higher level of integration within the subsystem than a level of integration that neurons within this subsystem have with neurons outside of the subsystem. In this context, high values of complexity reflect the coexistence of a high degree of functional integration, which form the "dynamic core" within a system (Tononi and Edelman 1998).

This concept of complexity can be applied to neurophysiological data and enables the degree to which neural processes are integrated and/or differentiated to be evaluated (Friston et al. 1995). It is also possible to compare the values of neural complexity in different cognitive and arousal states or empirically test the relationships between brain complexity and levels of conscious experience (Tononi and Edelman 1998). Because consciousness is related to a high level of functional interaction among neurons, it is possible to predict that the complexity of the dynamic core might correlate with the conscious state of the subject. For example, neural complexity is likely to be higher during waking and REM sleep than during the deep stages of slow-wave sleep, and it is possible to expect that complexity might be extremely low during generalized epileptic seizures (Tononi and Edelman 1998).

In this context, the concept of complexity provides a very useful and promising tool for consciousness research that might be of significant assistance in finding a scientific explanation for the specific biophysical processes related to the dynamic integration of large-scale information processing in the brain.

2.2 Mechanisms of Consciousness and Chaos Theory

The concept of dynamical chaos was first developed by the French mathematician, Henri Poincaré (1854–1912), who studied predictability in system behavior and found that chaotic pseudo-randomness is caused by high system sensitivity leading to disproportionate changes as a response to stimuli that influence system behavior (Poincaré 1908/1998; Peterson 1993). As a consequence, the sensitivity significantly decreases the ability to predict system behavior, which leads to information loss about later system development. In his work *Science and method* (p. 68) Poincaré (1908/1998) wrote: "A very small, unnoticeable cause can determine a visible very large effect; in this case we claim that this effect is a product of random .... However, even if the natural laws were perfectly known, we will never be able to know the initial conditions with some approximation. If this allows us to know the future with the same approximation that is all we want. We will say that the phenomenon is foreseeable, that it is governed by laws; however this is not always the case, it is possible that very small initial differences lead to very large ones in the final state ...."

Although the nonlinear mathematical approach to the "chaotic phenomena" and complexity in nature has its roots in Poincaré's work in the latter years of the nineteenth century, its application to the field of psychology and neuroscience is relatively new. The purpose of using this method is the understanding of relatively short periods in the behavior of a system that are extremely sensitive to very small changes (the "sensitivity to initial conditions"). At critical times, this sensitivity characterizes the initiation of new trends in the system's evolution, which may later emerge as very different macroscopic patterns of neural activity and mental processes (Elbert et al. 1994; Freeman 1983, 1991, 2000; Birbaumer et al. 1995; Kantz and Schreiber 1997; Meyer-Lindenberg et al. 2002; van Putten and Stam 2001; Faure and Korn 2001; Globus and Arpaia 1994; Korn and Faure 2003).

Several authors proposed that chaotic transitions might emerge in a wide variety of cognitive phenomena and might be linked to specific changes during the development of mental disorders, such as depression or schizophrenia (Pediaditakis 1992; Schmid 1991; Barton 1994; Gottschalk et al. 1995; Huber et al. 1999; Melancon and Joanette 2000; Korn and Faure 2003; Paulus and Braff 2003; Bob 2007; Bob et al. 2009a, 2009b), and might underlie psychological hypersensitivity to outside stimuli and the pathological processing of these stimuli.

Because of this sensitivity and unpredictability these nonlinear dynamical systems, although they might be deterministic, exhibit complex and random-like behavior. As experimental research indicates, values of the measured properties of many biological systems look random and their determinants are frequently unknown because of the high complexity of factors that influence the state of the living organism (Elbert et al. 1994; Freeman 2000; Dokoumetzidis et al. 2001; Korn and Faure 2003). The concept of randomness relies on evidence that every complex system has a large number of degrees of freedom that cannot be directly observed and manifest through the system's fluctuations (Elbert et al. 1994; Dokoumetzidis et al. 2001; Freeman 1991, 2000, 2001). Recent research shows that the chaotic, deterministic, dynamical systems display random-like behavior that is often indistinguishable from truly random processes (Elbert et al. 1994; Dokoumetzidis et al. 2001). However, there is evidence that chaotic dynamics tends to produce a spontaneous order and patterns of organization – self-organization (Elbert et al. 1994; Freeman 2001; Dokoumetzidis et al. 2001; Korn and Faure 2003). The self-organization patterns are typically linked to states of instability, which may result in new modes of behavior. The sudden phase transitions, called bifurcations, present a typical form of a system's behavior that is deterministic and characterized by typical modes of behavior called the "attractor." The attractor represents typical states of the system that describe its behavior, which is not random, and therefore it is compressed to a limited subset of all possible states in the "state space" (all the possible states that a system could have in principle) (Elbert et al. 1994; Freeman 2001; Dokoumetzidis et al. 2001). On the other hand, a random system has no restriction on its behavior ("it is random") and there is no limitation to its behavior in the "state space."

In this sense, it is possible to use the term state space for various phenomena, from the position of a particle to states of the human mind (in many cases it may be useful to perform a graphical representation of the state space). Deterministic systems are in their behavior strictly limited, and the resulting behavior, such as movement of a very small body (for example, in a gravitational field), is precisely defined and predictable, which means that under a constant condition (of gravitation) the body falls and then does not move. The behavior of such a system is strictly limited and defined in time and space (which together define the state space of the body). In this context, the state space can be visualized as a state space diagram, in which every possible state of the system corresponds to a unique point in the space. The number of dimensions or parameters of this space represents the degree of freedom of the system, and every dimension can be represented as an axis. Of course, in cases of complex behavior, it is difficult or impossible to imagine the state space, similar to the way in which we cannot imagine a cube or sphere that has more than three dimensions. Thus, the term multidimensional space presents an analogy to usual experience, which is used for the definition of mathematical terms. For example, the state space (which includes spatial and temporal dimensions) of the mechanical system can be described by all possible values of position and momentum or in the thermodynamic states or the phases of a chemical system, which may be described as a function of pressure, temperature or composition (Elbert et al. 1994; Dokoumetzidis et al. 2001).

What is specific to dynamical and chaotic systems is that they have limited behavior and a limited space of occurrence in the state space, similar to other deterministic systems, but their behavior has limited predictability, or rather it is unpredictable in space and time. A specific form of behavior of the chaotic system defined by the attractor includes the spatial and temporal dimensions of all its possible states in the past and future, which can be described as a "geometrical object" in the state space. In other words, this means that the dynamic and chaotic systems are neither deterministic nor random. Scientific description of a complex macrosystem, such as a living organism, may be defined by various complex "state functions," such as temperature, blood pressure, blood flow or electrical activity, for example, EEG, ECG, electrodermal activity (EDA), and other physiological, behavioral or cognitive characteristics (Freeman 1991, 2000; Elbert et al. 1994; Globus and Arpaia 1994; Gottschalk et al. 1995; Huber et al. 1999; Melancon and Joanette 2000; Faure and Korn 2001; Meyer-Lindenberg et al. 2002; Korn and Faure 2003; Paulus and Braff 2003; Breakspear 2006; Bob 2007; Bob et al. 2009a, 2009b).

Seminal contributions to this field of research were made by Walter Freeman, who was particularly interested in exploring how the brain generates cognitive processing, intentionality, and meaning. His main body of research was focused on the EEG study of perceptual processing in rabbits. In his research, Freeman found that activity in the olfactory cortex is chaotic, and proposed that chaos might underlie the basic forms of collective neural activity in perceptual processing, including the ability to access memorized sensory patterns and learn novel sensory information (Freeman 1991, 2000, 2001; Skarda and Freeman 1987). Freeman also proposed that chaos might explain the brain's ability to respond flexibly to the outside world and to generate novel activity patterns that are subjectively experienced as "novel" ideas, generated by unpredictable attractors that enable complex dynamic behavior of the brain and intentional behavior (Freeman 1991, 2000, 2001; Skarda and Freeman 1987). In this context, chaos theory enables understanding of the collective neural activity and brain functions as a global integrative process based on dynamic collections of attractors. These form an "attractor landscape" that is generated in the web of synaptic connections and represent behavioral "intentional" patterns that can be modified by learning (Skarda and Freeman 1987; Freeman 2000).

Within this framework Freeman (1999) proposed that the linear view of the stimulus–response reflex determinism is not an appropriate concept for behavioral dynamics and suggested that it might be needed to study behavioral responses and intentional behavior as consequences of nonlinear chains of various stimuli and responses. Freeman (1999, 2000) also suggested that the chaotic and complex self-organization of multilevel interactions between microscopic neurons in assemblies and the macroscopic emergent states is not possible within the concept of "linear causality," and must be replaced by "circular causality" (or reciprocal causality), which enables reflection upon the extensive relations among mutual dependencies, actions, and influences. Although the neurophysiological basis of these integrative processes is only partially understood, the concept of circular causality as a formal semantic description of brain dynamics related to chaotic self-organization and multimodal macroscopic patterns of neural activations may help to explain some of the functions of consciousness and intentional actions.

References

Atmanspacher H, Fach W. Acategoriality as mental instability. J Mind Behav. 2005;26:181–205.

Barton S. Chaos, self-organization, and psychology. Am Psychol. 1994;49:5–14.

Birbaumer N, Flor H, Lutzenberger W, Elbert T. Chaos and order in the human brain. Electroencephalogr Clin Neurophysiol Suppl. 1995;44:450–9.

Bob P. Hypnotic abreaction releases chaotic patterns of electrodermal activity during dissociation. Int J Clin Exp Hypn. 2007;55:435–56.

Bob P, Susta M, Gregusova A, Jasova D. Dissociation, cognitive conflict and nonlinear patterns of heart rate dynamics in patients with unipolar depression. Prog Neuropsychopharmacol Biol Psychiatry. 2009a;33:141–5.

Bob P, Susta M, Chladek J, Glaslova K, Palus M. Chaos in schizophrenia associations, reality or metaphor? Int J Psychophysiol. 2009b;73:179–85.

Breakspear M. The nonlinear theory of schizophrenia. Aust N Z J Psychiatry. 2006;40:20–35.

Bressler SL. Large-scale cortical networks and cognition. Brain Res Rev. 1995;20:288–304.

Crick F, Koch C. The problem of consciousness. Sci Am. 1992;267(3):153–9.

Crick F, Koch C. A framework for consciousness. Nat Neurosci. 2003;6:119–26.

Dennett D. Consciousness explained. Boston: Little, Brown; 1991.

Dokoumetzidis A, Iliadin A, Macheras P. Nonlinear dynamics and chaos theory: Concepts and applications relevant to pharmacodynamics. Pharmacol Res. 2001;18:415–26.

Duch W. Brain-inspired conscious computing architecture. J Mind Behav. 2005;26:1–21.

Edelman G. The remembered present. New York: Basic Books; 1989.

Elbert T, Ray WJ, Kowalik ZJ, Skinner JE, Graf KE, Birbaumer N. Chaos and physiology: deterministic chaos in excitable cell assemblies. Physiol Rev. 1994;74:1–47.

Faure P, Korn H. Is there chaos in the brain? I. Concepts of nonlinear dynamics and methods of investigation. C R Acad Sci III. 2001;324:773–93.

Frackowiak RSJ. Human brain function. San Diego, CA: Academic Press; 1997.

Freeman WJ. The physiological basis of mental images. Biol Psychiatry. 1983;18:1007–25.

Freeman WJ. The physiology of perception. Sci Am. 1991;264:78–85.

Freeman WJ. Consciousness, intentionality, and causality. J Conscious Stud. 1999;6:143–72.

Freeman WJ. Mesoscopics neurodynamics: from neuron to brain. J Physiol Paris. 2000;94: 303–22.
Freeman WJ. Biocomplexity: adaptive behavior in complex stochastic dynamical systems. Biosystems. 2001;59:109–23.
Friston KJ, Tononi G, Sporns O, Edelman GM. Characterising the complexity of neuronal interactions. Hum Brain Mapp. 1995;3:302–14.
Globus GC, Arpaia JP. Psychiatry and the new dynamics. Biol Psychiatry. 1994;32:352–64.
Gottschalk AM, Bauer MS, Whybrow PC. Evidence of chaotic mood variation in bipolar disorder. Arch Gen Psychiatry. 1995;52:947–59.
Huber MT, Braun HA, Krieg JC. Consequences of deterministic and random dynamics for the course of affective disorders. Biol Psychiatry. 1999;46:256–62.
James W. The principles of psychology. New York: Holt; 1890.
John ER. The neurophysics of consciousness. Brain Res Rev. 2002;39:1–28.
Kantz H, Schreiber T. Nonlinear time series analysis. Cambridge: Cambridge University Press; 1997.
Korn H, Faure P. Is there chaos in the brain? II. Experimental evidence and related models. C R Biol. 2003;326:787–840.
Libet B. Do the models offer testable proposals of brain functions for conscious experience. In: Jasper HH, Descarries L, Costelucci VC, Rossignol S, editors. Advances in neurology: consciousness at the frontiers of neuroscience. Philadelphia: Lippincott-Raven; 1998. p. 213–17.
Lumer ED, Edelman GM, Tononi G. Neural dynamics in a model of the thalamocortical system. I. Layers, loops and the emergence of fast synchronous rhythms. Cereb Cortex. 1997;7: 207–27.
Marcel AJ. Conscious and unconscious perception: an approach to the relations between phenomenal experience and perceptual processes. Cogn Psychol. 1983;15:238–300.
Melancon G, Joanette Y. Chaos, brain and cognition: toward a nonlinear order? Brain Cogn. 2000;42:33–6.
Merikle PM, Smilek D, Eastwood JD. Perception without awareness: perspectives from cognitive psychology. Cognition. 2001;79:115–34.
Meyer-Lindenberg A, Zeman U, Hajak G, Cohen L, Berman KF. Transitions between dynamical states of differing stability in the human brain. Proc Natl Acad Sci USA. 2002;99:10948–53.
Newman J. Thalamic contributions to attention and consciousness. Conscious Cogn. 1995;4:172–93.
Papoulis A. Probability, random variables, and stochastic processes. New York: McGraw-Hill; 1991.
Paulus MP, Braff DL. Chaos and Schizophrenia: does the method fit the madness? Biol Psychiatry. 2003;53:3–11.
Pediaditakis N. Deterministic non-linear chaos in brain function and borderline psychopathological phenomena. Med Hypotheses. 1992;39:67–72.
Peterson I. Newton's clock: Chaos in the solar system. New York: W.H. Freeman; 1993.
Picton TW, Stuss DT. Neurobiology of conscious experience. Curr Biol. 1994;4:256–65.
Poincaré H. Science and method. Londong: Thomas Nelson and Sons; 1908/1998.
Roland PE. Brain activation. New York: Wiley-Liss; 1993.
Schmid GB. Chaos theory and schizophrenia: elementary aspects. Psychopathology. 1991;24: 185–98.
Skarda CHA, Freeman WJ. How brains make chaos in order to make sense of the world. Behav Brain Sci. 1987;10:161–95.
Sporns O, Tononi G, Edelman GM. Connectivity and complexity: the relationship between neuroanatomy and brain dynamics. Neural Netw. 2000;13:909–22.
Sporns O, Tononi G, Edelman GM. Theoretical neuroanatomy and the connectivity of the cerebral cortex. Behav Brain Res. 2002;135:69–74.
Squires EJ. Why are quantum theorists interested in consciousness. In: Hameroff SR, Kaszriak A, Scott AC, editors. Toward a science of consciousness II: the second Tucson discussions and debates. Cambridge: MIT Press; 1998. p. 609–18.
Tononi G, Edelman GM. Consciousness and complexity. Science. 1998;282:1846–51.

Tononi G, Edelman GM. Schizophrenia and the mechanisms of conscious integration. Brain Res Rev. 2000;31:391–400.
Tononi G, Sporns O, Edelman GM. Reentry and the problem of integrating multiple cortical areas: simulation of dynamic integration in the visual system. Cereb Cortex. 1992;2:310–35.
Tononi G, Edelman GM, Sporns O. Complexity and coherency: integrating information in the brain. Trends Cogn Sci. 1998a;2:474–84.
Tononi G, McIntosh AR, Russell DP, Edelman GM. Functional clustering: identifying strongly interactive brain regions in neuroimaging data. Neuroimage. 1998b;7:133–49.
Van Putten MJAM, Stam CJ. Is the EEG really "chaotic" in hypsarrhythmia. IEEE Eng Med Biol Mag. 2001;20:72–9.
Velazquez JLP, Cortez MA, Snead III OC, Wennberg R. Dynamical regimes underlying epileptiform events: role of instabilities and bifurcations in brain activity. Physica D. 2003;186: 205–20.

Chapter 3
Consciousness and Functional Connectivity

This chapter describes recent findings indicating that the binding and synchronization of distributed neural activities that enable information integration are crucial to the mechanism of consciousness, and there is increased evidence that disrupted binding and information integration produce disintegration of consciousness in various neuropsychiatric disorders. These disturbed interactions produce patterns of temporal disorganization with decreased functional connectivity that may underlie specific perceptual and cognitive states. Together, these findings suggest that the process of neural or cognitive unbinding might influence more irregular neural states with higher complexity, and negatively affect information integration in the brain, which may cause disintegrated conscious experience, a decreased mental level or the loss of consciousness.

The process of cognitive binding refers to a specific integration of neural activities that enables subjective experiences. As previously reviewed data suggest, a basic mechanism of this binding process is the synchronization of functionally distinct cognitive modules (Van De Grind 2002).

This line of investigation opened up the possibility of understanding consciousness as a process of specific information integration. The starting point of this approach comprises clinical evidence that certain parts of the brain are more essential to consciousness than others, for example, the cerebral cortex contributes to different modalities and submodalities of consciousness more than other brain structures (Tononi 2005). Additionally, consciousness seems to be dependent on the temporal organization of neural activity. For example, consciousness is greatly reduced during slow wave sleep and generalized seizures, even though the levels of neural activity are comparable to or higher than those in wakefulness. Together, these findings suggest that although neural binding and large-scale synchronization might be important processes that enable consciousness, there is also the possibility that certain structures might be able to perform information integration better than others. Following these findings, the information integration concept of consciousness has been proposed, which suggests that consciousness might correspond to a certain capacity of the system to integrate information (Balduzzi and Tononi 2008). This information integration approach enables both neuroanatomical factors and

P. Bob, *Brain, Mind and Consciousness: Advances in Neuroscience Research*,
DOI 10.1007/978-1-4614-0436-1_3, © Springer Science+Business Media, LLC 2011

neurophysiological factors to be included that determine to what extent a neural structure can integrate information, and may also help to find out how certain information integration deficits may be related to various neuropsychiatric disorders.

3.1 Information Disintegration and Schizophrenia

According to recent evidence, mental disintegration in schizophrenia is related to disturbed binding and integration of multiple and disparate neural activities underlying cognitive brain functions and consciousness (Singer 1993, 2001; Varela et al. 2001; Fries 2005; Tononi and Edelman 2000; Peled 1999; Lee et al. 2003; Ford et al. 2007). These findings indicate an experimental paradigm for the understanding of conscious integration, which suggests conceptualization of schizophrenia as a disorder of neural integration (Tononi and Edelman 2000). The conceptualization of brain and cognitive disintegration in schizophrenia is reminiscent of the concept of "schizophrenia" proposed by Bleuler in his *Dementia Praecox*, or the group of the schizophrenias emphasizing the splitting or disintegration of consciousness (Bleuler 1911/1955; 1918/1906; 1924).

3.1.1 Information Disintegration in Schizophrenia and Corollary Discharges

A specific form of mental disintegration in schizophrenia is related to deficits in communication between the frontal and temporal lobes (Ford et al. 2005). According to recent evidence this disintegration of consciousness probably produces defective self-monitoring and self-experiencing (Feinberg 1978; Ford et al. 2001, 2007) and this lack of interaction and disintegration likely reflects the process of functional segregation of sets of neurons localized in different cortical areas. Experimental evidence for the relationship between conscious disintegration and defective self-monitoring or self-experiencing during hallucinations has been reported in several studies (Feinberg 1978; Ford et al. 2001, 2007; Poulet and Hedwig 2007). For example, using the PET scan it has been found that application of the same task to people with schizophrenia, and comparing hallucinators with nonhallucinators, shows that the hallucinators have decreased blood flow in the speech monitoring areas, such as the left middle temporal gyrus and the supplementary motor area (Andreasen 1997).

This loss of distinction between internally generated psychic activity and external input as a cause of hallucinations was proposed for the first time by Hughlings Jackson with an attempt to explain the "dreamy states" that occur in temporal lobe epilepsy (Meares 1999). In his concept, Jackson proposed that thinking may also be considered the highest and most complex motor activity and that defective self-monitoring and self-integrity originate in motor brain structures (Feinberg 1978;

Feinberg and Guazzelli 1999). Later findings indicate that motor commands from these brain structures are associated with neural discharges that alter activity in both sensory and motor pathways. These neural discharges, called corollary discharges (or efference copy), have a unique integrative function that enables monitoring and modification of the commands themselves before the effector event. In addition, they enable sensory systems to be informed that the stimulation produced by movement is self-generated or generated by an environment, which is crucial for the distinction between self and nonself (Feinberg 1978; Ford et al. 2001, 2005, 2007; Poulet and Hedwig 2007).

In addition, there is also evidence that the derangement of corollary discharges included in motor mechanisms of thinking produces many symptoms of schizophrenia in the visual or auditory system. For example, there is evidence that self-generated eye movements generate a "corollary discharge," or "efference copy" of the motor plan, informing the visual cortex that the changing of a visual input results from a self-generated action. A similar mechanism likely exists in the auditory system, where corollary discharges from motor speech commands prepare the auditory cortex for self-generated speech, perhaps through a link between frontal lobes, where speech is generated, and temporal lobes, where it is heard (Ford et al. 2001, 2007; Poulet and Hedwig 2007), for example, inner speech is misidentified as external voices (Ford et al. 2001, 2007; Poulet and Hedwig 2007).

Together these findings provide direct neurophysiological evidence for the corollary discharges that may transform sensory responses into self-generated ones and relative to externally presented percepts the processing of which fails in patients with schizophrenia in comparison to healthy subjects. In the context of binding and information integration current research findings on corollary discharges also present evidence that the process of disintegration in schizophrenia is related to defective communication between the structures of the frontal, temporal, and occipital lobes, which produces patterns of temporal disorganization and decreased functional connectivity.

3.1.2 *Neural Disintegration and Brain Complexity in Schizophrenia*

The findings reviewed above seem to be in agreement with the idea that mental disintegration could be a significant factor in the development of schizophrenia that might be related to disturbed neural integration (Tononi and Edelman 2000; Peled 1999; Lee et al. 2003; Bob et al. 2010a). According to recent evidence, the process of disturbed neural integration leading to increased functional segregation among groups of neurons might also be quantified using concepts from statistical information theory, and in particular by neural complexity, which measures the extent to which a pattern of functional connectivity is segregated or integrated (Sporns et al. 2000, 2002; Bob et al. 2009, 2010a). Increased functional segregation and complexity are often associated with symmetry-breaking and the ability of a system to have different states, which is also associated with a decrease in space coherence (Weng et al. 1999).

According to current findings, increased neural complexity observed from the EEG and other psychophysiological measures reflects processes during the activity of independent areas that enable fast parallel information processing running in a distributed mode (Klonowski et al. 1999; Sammer 1996; Elbert et al. 1994; Svetlak et al. 2010; Bob et al. 2009, 2010b). This means that numerous processes from sensory and cognitive channels are executed simultaneously and this desynchronized neural state may be related to active information processing in the cortex (Tirsch et al. 2004).

Together, these findings suggest that greater complexity related to more irregular neural states might influence neural and cognitive unbinding, and negatively affect synchronization phenomena in the brain, which are closely linked to an integrated conscious experience. Because the complexity may reflect a number of independent neural assemblies, it is possible that heightened complexity might be related to a number of independent clusters in associative chains or ideas due to disorganization in cognition and mental associations. These cognitive deficits may result in dissociative symptoms in schizophrenia and increased complexity in verbal associations. Increased neural complexity in the case of schizophrenia, therefore, may be a neural process that determines random-like and disorganized cognition, although in fact it may be characterized by complex, law-mediated behavior (Leroy et al. 2005). These processes may consequently lead to information overload, deficits in attentional filtering, and frontal lobe executive dysfunction (Hotchkiss and Harvey 1990; McGrath 1991; Goldberg and Weinberger 2000). This is compatible with the view that schizophrenia produces attentional modulation failure, which would be associated with a breakdown in the selective enhancement or inhibition of semantic representations whose underlying networks are widely distributed across the left (dominant) temporal and frontal lobes (Nestor et al. 1998, 2001). Supporting evidence, such as word recall studies, indicates disturbances in connectivity that are linked to associative strength. Furthermore, studies have shown the influence of NMDA receptor antagonists in recurrent inhibition, which produces a schizophrenia-like disturbance in association patterns and dysregulation in the suppression of associations (Nestor et al. 1998, 2001; McCarley et al. 1999). The dysregulation could be closely related to defective attentional filtering and a failure to inhibit activity of irrelevant neural assemblies (Vaitl et al. 2002; Olypher et al. 2006) and pathologically increased neural complexity (Bob et al. 2009). Such states may implicate the chaos that occurs when the neural process involves a large number of complex, interlinked, and simultaneously active states (Korn and Faure 2003). The increased complexity is linked to competing associations caused by the presence of a conflict among them that determines defective attentional filtering. Although these possibilities of explanation are very interesting and hopeful, it is necessary to note that the relationship among neural disintegration, dissociation, and schizophrenia described is clearly speculative, and further work is required.

A way of resolving the cognitive conflict is provided by discrimination among mental events in accordance with criteria for the interpretation of processed information (Baars 2002). Conflict monitoring is related to higher activation of the ACC

(Bunge et al. 2001; Paus 2001). The ACC is a part of the central network that also includes the insula and medial temporal lobe structures, such as the amygdala and the hippocampus, that integrate emotional and cognitive information and exert a modulatory effect on the lower brain centers that control the autonomic nervous system and modulate autonomic responses (Benarroch 1993; Critchley 2002). In this way, lasting psychological conflict in schizophrenia patients could be linked to disruptions in the integrity of the limbic system network, particularly the cingulate gyrus, that play a significant role in the pathophysiology and psychopathology of schizophrenia, as suggested by brain imaging studies (Fujiwara et al. 2007). These links indicate that the increased complexity in neural activity in schizophrenia patients could be related to the inner conflict, which is subjectively experienced as "inner chaos."

3.2 Epileptiform Processes and Information Integration

Epileptic seizure is characterized by transient signs and/or symptoms of abnormal, excessive or synchronous neuronal activity in the brain (Fisher et al. 2005). This process is generally explained by strongly connected networks related to neural bursting as a prerequisite for the generation of synchronized neural activity and seizure-like bursting, but, on the other hand, weakly coupled cortical networks and a reduction in synaptic transmission can create the same process (van Drongelen et al. 2005). The seizure-like bursting may result in a seizure, but may also proceed in the form of seizure-like conditions on a subclinical level without the presence of a neurologically defined seizure. This issue is terminologically defined by a clear distinction in the relationship between epileptic and epileptiform discharges, which describes the connection between ictal and interictal symptomatology. While epilepsy is defined as a chronic condition characterized by spontaneous, recurrent seizures, and a seizure is defined as a clinical event associated with a transient, hypersynchronous neuronal discharge (epileptic is a descriptive term used to denote the presence of epilepsy), epileptiform discharge is an interpretative term used in electroencephalography that applies to distinctive waves or complexes that are distinguishable from the background activity, resembling the waveforms recorded in a proportion of human subjects suffering from an epileptic disorder (Chatrian et al. 1974). Epileptiform patterns include spikes and sharp waves, alone or accompanied by slow waves, occurring singly or in bursts lasting at most a few seconds. The term epileptiform typically refers to interictal paroxysmal activity and not to the EEG activity seen during an actual seizure, which is called an electrographic seizure (Chatrian et al. 1974).

In agreement with the binding and information integration concept of consciousness and cognitive processes, changes in neural connectivity and synchrony not only relate to epileptic seizures, but also, in epileptiform conditions, significantly affect the content of consciousness, and may lead to its loss, mainly in cases of generalized seizures. This evidence suggests that various conditions leading to changes in consciousness or loss of consciousness might be in agreement with the basic concept of binding, which, by means of neural synchrony brings together processed information and creates subjective experience and consciousness. The concept of binding and information integration in neuropsychiatric disorders is also in agreement with recent evidence in anesthesiology, indicating that the anesthetic-mediated interruption of consciousness may relate to an "unbinding" mechanism across various parts of the brain (Mashour 2004), and that a common neurophysiology may underlie the loss of consciousness or a diminished level of consciousness in schizophrenia and other psychiatric disorders (Bob et al. 2010a). An interruption of information integration (Lee et al. 2009a,b), as well as the disruption of the frontoparietal network (Lee et al. 2009a,b), has been associated with the induction of general anesthesia, despite the fact that isolated information processing can still persist. Mashour (2008) has suggested that cognitive unbinding might be a common mechanism for the interruption of normal consciousness seen in both anesthesiology and psychology, which may also potentially help to explain changes in consciousness in various neuropsychiatric conditions.

3.2.1 Epileptiform Activity and the Neural Correlate of Consciousness

The historical background for research into the relationship between epileptiform activity and the neural correlate of consciousness originates in the works of Hughlings Jackson, who studied the dreamy states related to temporal lobe seizures. In his writings, Hughlings Jackson described in detail focal motor seizures, in particular focal seizures of temporal lobe origin, which he called the dreamy states, including the implications of these seizures for the understanding of basic mechanisms of consciousness (Hughlings Jackson 1931; Meares 1999; Hogan and Kaiboriboon 2003, 2004). Hughlings Jackson's descriptions of the dreamy states first appeared in the 1870s and included typical symptoms such as hallucinations, strangeness, the unreality of things, such as derealization or depersonalization, double consciousness (looking at one's self through the eyes of others), *déjà vu* and *jamais vu*, various autobiographical memories including flashbacks, and other symptoms currently described in medial temporal lobe epilepsy. These symptoms occur as a consequence of epileptic discharges in the medial temporal lobe, mainly including limbic structures, such as the hippocampus and amygdala, that are involved in memory and emotional processing (Eadie 1998; Teicher et al. 1993, 2006a).

Typical examples of these symptoms are included in the self-reported questionnaire, the Limbic System Checklist (LSCL-33; Teicher et al. 1993), which includes typical symptoms that occur without outside stimulation, like headache, numbness or tingling, dizziness, the sensation of something crawling under the skin, a flushing or hot sensation, feeling that the heart has stopped or is pounding or racing, flashing lights, hearing a voice calling the subject's name, hearing a voice, repeating a sentence or phrase, memory flashbacks (for example, feeling exactly as you did as a child during an upsetting event), sensations that events, conversations, or places are strangely familiar, as if you had experienced or dreamed the situation before, the sensation that a familiar person or place has become unfamiliar, changed, different, or almost as if you had never experienced them or it before.

Hughlings Jackson also described interpretation of the dreamy states as a release of the lower level of brain functions because of inactivated inhibition of the highest functional levels caused by the epileptic discharges, and not as a direct consequence of the epileptic discharges. This interpretation of the dreamy states is in agreement with the proposed concept of functional deactivation of the amygdala and hippocampus by epileptiform activity, leading to liberation of the neocortical structures (Halgren et al. 1978; Bancaud et al. 1994). This interpretation seems to be reasonable, but in the context of binding related to neural synchrony, there is also the possibility that epileptiform or epileptic synchronization might produce unusual contents of consciousness directly due to an abnormal level of binding and information integration caused by epileptiform synchronization. This second hypothesis is in agreement with the interpretation proposed by Gloor et al. (1982), who considered the dreamy states to be a positive phenomenon linked to the activation of medial temporal lobe structures. In agreement with this hypothesis, there is evidence from neuroimaging data that indicates active involvement of the hippocampus and other medial temporal lobe structures in memory activation (Nadel et al. 2000; Ryan et al. 2001), which supports the view of unusual integration of associations by a different mechanism of synchrony during activation of these structures (and not their deactivation).

From the point of view of the second hypothesis, the integrative process related to abnormal epileptic synchrony enables the simultaneous activation of neuronal clusters in a different manner than that usually used during normal brain states. These transient periods of hypersynchronized firing create a different and "unusual" pattern of a neural coherent whole that enables various memories and associations to be connected in a different way than that during the normal waking state, which produces the dreamy states. Consequently, these different levels of binding and information integration lead to different methods of perception and cognitive processing, which significantly change the contents and subjective quality of consciousness in a similar way to that in dreams, in which the usual order of cognitive processing is substituted by the unusual or "strange" order of the dream. In this context, it is possible to understand states of consciousness as a continuum (or spectrum) of various levels of binding between unconsciousness, such as, for example, in anesthesia (unbinding), and higher forms of consciousness during vigilant states, including altered states of consciousness that may occur in hypnosis, sensory deprivation, meditation, states of consciousness affected by drugs or states of consciousness in psychiatric disorders. From this point of view the states of consciousness can be explained by different levels of binding and information integration of simultaneously active neuronal assemblies into a specific coherent whole. This extremely complex process of binding and integration may globally operate analogous to a process of interference that may produce a meaningful pattern (positive interference) or no pattern caused by negative interference. This interpretation seems to be reasonable, mainly in light of the current findings that neural activity during seizures is not a random process, but is pseudo-random activity that represents an abnormal level of complex neural activities and self-organization.

In his writings Hughlings Jackson reported numerous cases of the "dreamy states," and described a close association between the "dreamy states" and the immediate onset of epileptic discharges (Hughlings Jackson 1931). A new era in "dreamy states" research started with discoveries of experimental electrical brain stimulation. Seminal works in this research field were reported by Penfield, who found that visual hallucinations were elicited by electrical stimulation in sites widely distributed over the temporal neocortex (Penfield 1967; Penfield and Perot 1963). Penfield reported that these experiential hallucinations, responding to past experiences, reflected a bi-directional activation of the temporal neocortex, which is involved in memory storage, and the centrencephalic system, which participates in the integration of memories. Further research has shown that the dreamy states may be elicited by stimulation of the medial temporal lobe structures (Ferguson et al. 1969; Halgren et al. 1978; Gloor et al. 1982; Gloor 1990; Bancaud et al. 1994). In 2007, an interesting series of cases was reported by Vignal et al., who studied spontaneous and provoked dreamy states using electrical discharges localized within medial temporal lobe structures, and found that the early spread of the discharges to the temporal neocortex prevented the occurrence of the dreamy state. They found a clinical continuity between experiences that had already been lived (*déjà vécu*) and visual hallucinations. The visual hallucinations consisted of personal memories that were recent, distant, or from childhood. Vignal et al. (2007) found that these memories that were relived by the subject, with only one exception, differed from one seizure to another, but were always from the same period of life. In agreement with further evidence emphasizing the central role of the amygdala and hippocampus (both left and right) in the recall of recent and distant memories, they also found that the pathological activation of the amygdala and hippocampus during seizures may trigger memory recall.

A detailed analysis of psychiatric presentations observed in patients with temporal lobe epilepsy was reported by Ferguson et al. (1969), who found that psychiatric manifestations in these patients were related to disturbances in higher cortical functions and that psychiatric symptomatology apparently varied with the seizure type. They also found that the interindividual variations in psychotic symptomatology were significantly related to events in the patient's past and current emotional life, which reflected the form and content of psychotic experiences. In agreement with their findings Ferguson et al. (1969) hypothesized that a local brain lesion might produce a fixed anatomical diminution of neural elements and/or temporary functional reduction in the activity of available neural elements by ictal discharge. They suggested that an increase in normal afferences to altered synaptic relays might lead to "overload" of the reduced capacity for excitation control and induce excessive output of neuronal activity in the form of epileptic discharges that disrupt the usual ability to integrate complex experiences. Ferguson et al. (1969) further suggested that the development of psychosis might depend on the combined activation of the medial and lateral structures of one or both temporal lobes with disruption of contact with reality by interference with the retention, recall, continuity, ordering, and integration of ongoing experience, which are functions essential for the ego (Ferguson et al. 1969).

In the context of these findings limbic kindling has been suggested as a model of epileptogenesis in focal human limbic epilepsy or complex partial seizures with secondary generalization (Adamec 1990, 1997; Albright and Burnham 1980; Loscher et al. 1986). This conceptual explanation is supported by findings of repeated electrical stimulation in the human hippocampus and thalamus, causing an epileptic disorder that had not been documented before the experiment (Adamec 1990, 1997; Sramka et al. 1977; Monroe 1982). Important evidence indicates reported data documenting the time-dependent spread of epileptic excitability independently on tissue pathology as a consequence of organic damage (Adamec 1990, 1997; Jensen and Baram 2000). Several data also suggest a delay between the trauma and the onset of seizures (Adamec 1990), and it has been found that successful prophylactic anticonvulsant therapy following head trauma with neurological signs of brain damage reduces the incidence of the development of an epileptic disorder (Adamec 1990, 1997; Servit and Musil 1981). The brain damage is likely to induce pathological discharges in affected parts of the neural network, which may lead to repeated stimulation. This phenomenon, known as "kindling," was first reported by Sevillano in 1961 and was later elaborated by Goddard, who comprehensively described using implants stereotactically implanted into the amygdala and other brain regions (Kraus 2000). In his research, Goddard found that repetitive subthreshold stimuli elicit only small changes in EEG activity or behavior, whereas stimuli applied later in a state of increased sensitivity result in local after-discharges, increased synchronization, or seizures (Goddard 1967; McNamara et al. 1992; Kraus 2000). In this context, the concept of kindling may also be used as an explanation of the psychopathological processes that may mirror altered limbic functions (Adamec 1990, 1997).

This close relationship between seizures and psychopathology is historically linked to the concept of biological antagonism between epilepsy and psychosis proposed by László von Meduna (Meduna 1934; Wolf and Trimble 1985; Krishnamoorthy et al. 2002). Meduna followed the study by Steiner and Strauss (Wolf and Trimble 1985), who investigated 6,000 schizophrenic patients and found typical epileptic paroxysms to be very rare. In his later work, published in 1935, Meduna reported a study of 176 patients of whom 95 with epilepsy had psychotic symptoms at the same time; he also found that a comorbid condition presenting both epileptic and psychotic symptoms may exist (Meduna 1935). In his medical practice Meduna used convulsive drugs (for example, camphor or pentetrazol) in the treatment of schizophrenia, and found that these drugs often cause convulsions and a lessening of schizophrenic symptoms (Wolf and Trimble 1985). Later discussions following Meduna's works continued the development of convulsive therapy in psychiatry.

Further evidence about the relationship between epilepsy and psychosis was reported in the 1950s by Heinrich Landolt, the director of a Swiss asylum for epileptics in Zürich, who documented findings on forced normalization using electroencephalography (Landolt 1953; Wolf and Trimble 1985; Faber and Vladyka 1987; Pakalnis 1988). Landolt introduced the term "forced normalization" to describe a specific defense mechanism of the brain that tends to decrease the frequency of epileptic discharges (Wolf and Trimble 1985). According to his findings this defensive response may lead to psychotic symptoms that can appear spontaneously or that may be related to antiepileptic medication. Landolt (1953) documented the phenomenon of forced normalization in patients with temporal lobe epilepsy and later also in patients with focal cortical epileptic seizures. A basic clinical characteristic of forced normalization is the development of acute psychotic states in patients without a previous psychiatric history, which is closely related to a decrease in epileptic EEG abnormalities and improvement in the controlling of seizures (Wolf and Trimble 1985). The clinical manifestations of forced normalization may also include dysphoric states, hysteria and hypochondria, affective disorders, and miscellanea (twilight states). Forced normalization can be observed in both generalized and partial epilepsy as a rare complication, but it is relatively frequently observed in adults with persistent absence seizures (Wolf 1991; Kanner 2000, 2001; Schmitz et al. 1999; Marsh and Rao 2002). In later studies forced normalization was also reported to be a response to the neurosurgery on the epileptic focus (Mace and Trimble 1991; Blumer et al. 1998), which may cause affective disorders or psychotic episodes due to the sharply reduced number of seizures postsurgery (Jobe et al. 1999; Kanner and Balabanov 2002). Several clinical data also suggest that antidepressant therapy might be crucial for a significant number of patients after surgical treatment of epilepsy and may prevent the symptoms of interictal dysphoric disorder (Jobe et al. 1999; Kanner and Balabanov 2002).

Several findings also indicate the existence of forced normalization between seizures and affective disorders (Jobe et al. 1999; Kanner and Balabanov 2002). The first reason is an increasing occurrence of depressive symptoms when the frequency of seizures decreases, and the second is that electroconvulsive therapy and chemically induced seizures have a high degree of efficacy in the treatment of depression and manic states (Jobe et al. 1999). In addition, there is growing evidence that antidepressant medication exhibits anticonvulsant effects and the use of these drugs often constitutes a safe therapeutic approach in epileptic patients with interictal dysphoric disorder (Chaplin et al. 1990; Trimble 1996; Jobe et al. 1999). According to recent clinical evidence, antidepressant drugs suppress seizures when the blood and brain concentrations of them are relatively low. Conversely, seizures may occur as a response to antidepressants in overdoses or in response to their excessive blood levels (Jobe et al. 1999). On the other hand, there is evidence that anticonvulsant medications have emerged as powerful agents for the treatment of bipolar disorders, schizoaffective disorder, or for the treatment of refractory depression (Jobe et al. 1999; Chaplin et al. 1990; Trimble 1996).

Because the term "forced normalization" has been defined in connection with EEG, Tellenbach, in the 1960s, introduced the concept of alternative psychosis, implying that stopping seizures does not mean the disappearance or inactivity of the pathological state (Wolf and Trimble 1985; Krishnamoorthy et al. 2002), and that in patients who display alternative psychosis, subcortical epileptic discharges, particularly in the limbic system, are continuously present (Wolf and Trimble 1985).

Robert Heath, in the 1960s, proposed an interesting interpretation of forced normalization based on arguments of clinical and experimental data (Heath 1961/2005).

His conceptual proposal, based on clinical and experimental EEG data, suggests that the syndrome of epilepsy might vary and not be adequately defined. According to Heath, the most consistent symptom of epilepsy comprises alterations of consciousness that are many and diverse. Heath (1961/2005) documented unique data obtained by depth EEG recordings of patients with seizures and schizophrenia, and found significant differences between the seizure group and the schizophrenic group during periods when the epileptic patients displayed clinical features indistinguishable from those of the schizophrenic patients. He found that both groups had EEG abnormalities in the hippocampus, amygdala and in the septal region. However, comparative analysis has shown that EEG abnormalities are more pronounced in the hippocampus and amygdala in epileptic patients and less pronounced in the septal region, whereas EEG abnormalities in the schizophrenic patients are located predominantly in the septal region. In the author's opinion, there is no reason to conclude that schizophrenia and epilepsy constitute a single disease just because the same anatomical structures are involved (Heath 1961/2005).

These findings suggest that although a significant overlap exists between epilepsy and several psychiatric disorders, it might be possible to find specific criteria for diagnostics and treatment. On the other hand, these data indicating a close relationship between psychiatric disorders and epilepsy strongly suggest that there might be a possibility of finding a common platform for understanding them. From these findings it is possible to assume that epilepsy and epileptiform activity are related to a neural correlate of consciousness, and that specific states of consciousness are determined by various lesions that may occur in the brain. These lesions likely have various causes that determine the pathological process, for example, psychotic behavioral symptoms and altered electrical activity are also present with tumors, degenerative processes, infections, and toxic agents affecting these structures (Heath 1961/2005; Bauer, and Bien 2009; Neligan and Shorvon 2010; Avila and Graber 2010). Several recent data suggest that epileptiform activity might also be caused by stress and psychological trauma, which significantly affect brain functions and alter neural excitability (Teicher et al. 2003, 2006a).

All these pathological changes specifically influence epileptic or epileptiform synchronization by affecting neurons in some specific structures of the brain and in the brain network as a whole. Because of these pathogenic influences some parts of the brain may be damaged, which consequently changes neural excitability and typically affects neural binding and information integration. This pathological state of binding consequently produces neural patterns of brain activity that create disease-specific cognitive patterns of perception, memory, and consciousness.

3.2.2 Sensitization, Kindling, and Epileptiform Changes in Schizophrenia

Recent evidence indicates that various environmental factors, such as perinatal damage, hypoglycemia, stressful experiences in childhood, and other influences interacting on a genetic basis, may significantly affect persisting sensitivity to stress and other stimuli at a later time (van Winkel et al. 2008; Eack et al. 2008; van Os et al. 2008). This state of increased sensitivity may be related to an imbalance in interactions between the dopaminergic and glutamatergic systems, altered dopamine neurotransmission, and the consequent alterations in cognitive biases that result in critical conditions in the pathogenesis of schizophrenia (Collip et al. 2008; Yuii et al. 2007). According to recent findings this increased sensitivity and imbalance of neural systems could be explained by sensitization, leading to an increased response to a certain stimulus, which at the beginning of this process was subthreshold and caused no or little response (Castner and Williams 2007; van Winkel et al. 2008).

A specific form of sensitization that might play a role in the pathogenesis of schizophrenia is the progressively increasing response of groups of neurons due to repetitive subthreshold stimulation—kindling—which later may lead to epileptic or epileptiform activity (Smith and Darlington 1996; Glenthoj and Hemmingsen 1997; Kraus 2000; Grossman et al. 2003). Recent findings suggest that the kindling related to focal after-discharges within the amygdala and other brain structures might cause local changes in synchronization and seizure-like activity that could be important in the pathogenesis of schizophrenia (Glenthoj and Hemmingsen 1997; Stevens 1999; Grossman et al. 2003).

These findings may potentially explain treatment resistance to usual antipsychotic medication in several schizophrenia patients and also the clinical importance of an appropriate anticonvulsant medication, even in patients who do not display seizures or epileptiform abnormalities on scalp EEG (Johannessen 2008; Tiihonen et al. 2009). Several findings also show that epileptiform activity can occur with indications similar to those of temporal lobe epilepsy, such as somatic, sensory, behavioral, and memory symptoms that may also occur in nonepileptic conditions (Roberts et al. 1992; Teicher et al. 2003; Bob 2003a). These symptoms have been found to play a role in schizophrenia too, and the significant presence of these symptoms in treatment-resistant patients might indicate a good response to anticonvulsant drugs (Bob et al. 2010b).

With respect to the specific changes in neural synchronization found in epilepsy, there is an open question for further research asking how the long-term effect of sensitization and kindling influences the microstructure and dynamics of the nervous system, which is characterized by increased neural excitability and hypersynchronized activity in response to an initially subthreshold stimulus. In a previous study (Bob et al. 2010b), we found preliminary evidence that an increase in the local EEG synchronization is specifically related to the content of consciousness in the form of symptoms, similar to those of temporal lobe epilepsy, as measured by the Limbic System Checklist (LSCL-33) (Teicher et al. 1993), which are experienced as somatic, sensory, behavioral, and memory symptoms. These symptoms may be generally described as brief hallucinations, paroxysmal somatic disturbances, and movement or psychological automatisms.

Taken together there is evidence that sensitization, or "kindling-like," phenomena, play an important role in the pathogenesis of schizophrenia and other mental disorders, and these have received a great deal of attention with regard to efforts to conceptualize the pathophysiology of seemingly diverse psychiatric disorders such as schizophrenia, mood disorders, drug addiction, or post-traumatic stress disorder (Collip et al. 2008; Kraus 2000; Phillips et al. 1997; Post et al. 1997; Post and Weiss 1996).

Within this context the dopaminergic hypothesis of schizophrenia also provides results that show that positive schizophrenic symptoms are consequences of hyper-dopaminergic kindling in the mesolimbic dopaminergic system (Adamec 1990; Glenthoj 1995; Stevens 1999; Grossman et al. 2003). The concept of kindling as a model for psychopathology in several schizophrenia patients is in agreement with recent findings that schizophrenia is often related to a loss of physiological balance between excitation and inhibition (Stevens 1999). Typical of this imbalance is that the normal equilibrium between excitation and inhibition is permanently altered by repeated focal excitation or kindling, resulting in a permanent state of excessive focal excitability and spontaneous seizures (Goddard 1967; Stevens 1999). Several recent findings suggest that similar kindling or sensitization might originate in inhibitory systems in response to focal physiological pulsed discharges of limbic neurons and that this excess of inhibitory factors might then manifest as a psychosis (Stevens 1999). This excessive focal inhibition may be induced by the increased release or increased receptor density of several inhibitory transmitters (Stevens 2002).

According to these findings discharges related to increased excitatory neural activity may also be modulated by a regionally specific compensatory upregulation of GABA-A receptors in response to decreased GABAergic input in hippocampal pyramidal cells (Heckers and Konradi 2002; Möhler 2006). In general, GABAergic neurons provide both inhibitory and disinhibitory modulation of cortical and hippocampal circuits, contribute to the generation of oscillatory rhythms, and participate in discriminative information processing that is typically affected in schizophrenia, such as the gating of sensory information and attentional filtering within the corticolimbic system (Glenthoj and Hemmingsen 1997; Benes and Berretta 2001; Costa et al. 2004; Gonzalez-Burgos and Lewis 2008; Jacob et al. 2008). In agreement with this role of GABAergic neurons in cognitive functions several findings also suggest that disturbances in the GABA system might be related to stressful conditions and alterations in the dopamine system (Benes and Berretta 2001; Teicher et al. 2003, 2006a; Yuii et al. 2007). Furthermore, an increased flow of excitatory activity from the basolateral nucleus of the amygdala may be related to disturbances in the GABA system (Benes and Berretta 2001).

Taken together, current findings suggest that there might be a link between disturbances in the GABA system and increased dopaminergic activity that may determine the relationship between reported symptoms similar to those of temporal epilepsy, reflecting the abnormal neural excitability that specifically influences the contents of consciousness and subjective experience.

3.2.3 Sensitization, Kindling, and Epileptiform Changes in Depression

Recent evidence indicates that the sensitization process with kindling-like progression frequently leads to maladaptive responses and negative influences on brain structures that manifest as an increased probability of relapses, recurrences, residual symptomatology, and other forms of psychopathology (Post 1992; Roberts et al. 1992; Segal et al. 1996; Teicher et al. 2003, 2006a). These changes may result in deficits in inhibitory functions and limbic system irritability manifesting as markedly increased prevalence of seizure-like symptoms that are similar to the psychosensory symptoms of temporal lobe epilepsy (Teicher et al. 1993; Teicher et al. 2003, 2006a). These findings are in agreement with evidence of a positive clinical response to anticonvulsant treatment in many depressive and other psychiatric patients, although the lack of clear nonbehavioral evidence of CNS dysfunction measured by EEG may frequently cover up the underlying neurological nature of these symptoms (Bob 2003a; Johannessen 2008; Post et al. 1988; Roberts et al. 1992; Silberman et al. 1985; Teicher et al. 2003, 1993; Varney et al. 1993). In addition, increased vulnerability related to sensitization and kindling may also cause the brain to become more sensitized to the depressive state and the onset of future episodes is less related to stressful life events than at the beginning of the disease (Keller 2003; Monroe and Harkness 2005). These reported findings are also in agreement with clinical evidence that many patients with unipolar depression are unresponsive to antidepressant treatment and that antiepileptic drugs may be an effective adjunctive treatment (Gabriel 2006; Johannessen 2008; Morishita 2009; Silberman et al. 1985; Varney et al. 1993; Vigo and Baldessarini 2009). For example, Silberman et al. (1985) assessed the occurrence of transient sensory, cognitive, and affective changes resembling those described by temporal lobe epileptic patients in 44 patients with affective illness, 37 with complex partial seizures, and 30 controls. Their results indicate that the symptoms occurred frequently in association with episodes of affective illness and epilepsy, but were rare in controls. They also reported that greater numbers of symptoms were associated with better response to lithium and tricyclic antidepressants. The authors conclude that transient sensory, cognitive, and affective phenomena may be more common in affective illness and other psychiatric conditions than is generally recognized, and may be clues to the underlying pathophysiology of these conditions (Silberman et al. 1985). Similarly, Varney et al. (1993), in a study of 13 depressed patients with documented histories of failure to respond to tricyclic antidepressant medications, also reported multiple partial seizure-like symptoms, and found that 11 of the 13 patients showed moderate or substantial improvement in affective status in response to carbamazepine. In addition, the mean number of reported partial seizure-like symptoms decreased significantly with treatment. The authors conclude that these preliminary observations show that there is likely to be a subgroup of treatment-resistant, carbamazepine-responsive depressive patients, who can be identified by evaluating for the presence of the seizure-like symptoms (Varney et al. 1993). In agreement with these findings our previous clinical study

(Bob et al. 2010d) in 113 patients with unipolar depression also indicates that in depressive patients seizure-like symptoms display a significant correlation with depression and specifically influence the contents of consciousness and subjective experience.

3.2.4 *Traumatic Stress, Sensitization, and Epileptiform Activity*

Increasing evidence in recent reports indicates that child abuse and other traumatic stress experiences caused by inescapable adverse physical, emotional or social events constitute significant conditions in the pathophysiology of various psychiatric disorders (Teicher et al. 2003; Duman and Monteggia 2006; Dranovsky and Hen 2006; Bogdan and Pizzagalli 2006; Cole et al. 2006; Harkness et al. 2006; Monroe et al. 2007). Early stress may determine developmental abnormalities in the amygdala, hippocampus, cerebellum, ACC, corpus callosum, and other brain structures that play a critical role in mediating response to stress (Teicher et al. 2003, 2006a; Riklan et al. 1977; Putnam 1995, 1997; Bremner 2006). Stress also significantly influences the decrease in brain-derived neurotrophic factor (BDNF) in the hippocampus, which may determine the neurodegenerative process (Duman and Monteggia 2006). Repeated stressful events may also determine sensitization, leading to an increase in responsiveness to a stress stimulus resulting from repeated stressors, with significantly increased vulnerability to stressors that have more lasting consequences with kindling-like progression (Post et al. 1995; Post and Weiss 1998; Kraus 2000). The kindling model of stress-related sensitization (Post et al. 1995) seems to be in agreement with suggestive evidence that stress can influence the significantly increased occurrence of EEG abnormalities that have been reported in considerably traumatized patients, mainly in the frontotemporal region. These abnormalities consisted of spikes, sharp waves, or paroxysmal slowing, predominantly in the left hemisphere (Teicher et al. 1993, 2003, 2006a; Putnam 1997; Ito et al. 1993). It has been proposed that stress-related sensitization might cause changes in GABA postsynaptic receptors, which may lead to overstimulation of neurons, mainly in the limbic system, resulting in limbic system irritability manifesting as markedly increased prevalence of symptoms suggestive of temporal lobe epilepsy (Teicher et al. 2003, 2006a; Post et al. 1995). Recent data strongly suggest that early stress might determine limbic irritability and temporolimbic, seizure-like activity (Teicher et al. 2003, 2006a; Spigelman et al. 2002), and a close link between limbic irritability and cerebellar vermis has been reported (Teicher et al. 2003, 2006a; Anderson et al. 2002). Anderson et al. (2002) found a reciprocal relationship between activity in the cerebellar vermis assessed by fMRI T2 relaxometry and symptoms of limbic irritability measured by the questionnaire LSCL-33 (Limbic System Checklist). They reported a significant correlation between T2 relaxation time and the degree of limbic irritability (LSCL-33) in healthy young adult controls ($r=-0.67$) and also in young adults who had a history of repeated sexual abuse ($r=-0.71$). Teicher et al. (1993, 2003, 2006a) also found that adult outpatients with a self-reported history of physical or sexual abuse had increased LSCL-33 scores, which were dramatically elevated in patients with a history of combined abuse, both physical and sexual. The results are consistent with findings that the cerebellar vermis controls limbic activation and inhibition, and influences the onset and spread of seizures (Heath 1976; Riklan et al. 1976; Schmahmann and Sherman 1998; Teicher et al. 2003, 2006a; Schutter and van Honk 2006). These findings suggest that cognitive and emotional dysregulation related to traumatic stress might be linked to defective inhibitory functions that may also lead to temporolimbic, seizure-like activity. This epileptic-like process may emerge in the form of symptoms similar to those of ictal temporal lobe epilepsy, such as somatic, sensory, behavioral, and memory symptoms that may also occur in nonepileptic conditions (Teicher et al. 2003, 2006a; Silberman et al. 1985; Roberts et al. 1992; Hines et al. 1995).

Taken together, current findings suggest that there might be a link between the influence of traumatic stress and psychosensory symptoms related to limbic irritability, which are similar to symptoms of temporal epilepsy and likely reflect abnormal neural excitability and disturbances in brain integrative processing that may specifically influence the neural correlate of consciousness and subjective experience.

References

Adamec R. Transmitter systems involved in neural plasticity underlying increased anxiety and defense-implications for understanding anxiety following traumatic stress. Neurosci Biobehav Rev. 1997;21:755–65.

Adamec RE. Does kindling model anything clinically relevant? Biol Psychiatry. 1990;27:249–79.

Albright PS, Burnham BM. Development of a new pharmacological seizure model: effects of anticonvulsants on cortical and amygdala-kindled seizures in the rats. Epilepsia. 1980;21: 681–9.

Anderson CM, Teicher MH, Polcari A, Renshaw PF. Abnormal T2 relaxation time in the cerebellar vermis of adults sexually abused in childhood: potential role of the vermis in stress-enhanced risk for drug abuse. Psychoneuroendocrinology. 2002;27:231–44.

Andreasen NC. Linking mind and brain in the study of mental illnesses: a project for a scientific psychopathology. Science. 1997;275:1586–93.

Avila EK, Graber J. Seizures and epilepsy in cancer patients. Curr Neurol Neurosci Rep. 2010;10: 60–7.

Baars BJ. The conscious access hypothesis: origins and recent evidence. Trends Cogn Sci. 2002;6: 47–52.

Balduzzi D, Tononi G. Integrated information in discrete dynamical systems: motivation and theoretical framework. PLoS Comput Biol. 2008;4(6):e1000091.

Bancaud J, Brunet-Bourgin F, Chauvel P, Halgren E. Anatomical origin of déjà vu and vivid 'memories' in human temporal lobe epilepsy. Brain. 1994;117:71–90.

Bauer J, Bien CG. Encephalitis and epilepsy. Semin Immunopathol. 2009;31:537–44.

Benarroch EE. The central autonomic network — functional organization, dysfunction and perspective. Mayo Clin Proc. 1993;68:988–1001.

Benes FM, Berretta S. GABAergic interneurons: implications for understanding schizophrenia and bipolar disorder. Neuropsychopharmacology. 2001;25:1–27.

Bleuler E. Dementia praecox or the group of the schizophrenias. New York: International University Press; 1911/1955.

Bleuler E. Consciousness and association. Eder MD (trans.). In: Jung CG, editor. Studies in word-association. London: William Heinemann; 1918/1906. p. 266–96

Bleuler E. Textbook of psychiatry. Brill AA (trans.). New York: Macmillan Publishing; 1924

Blumer D, Wakhlu S, Davies K, Hermann B. Psychiatric outcome of temporal lobectomy for epilepsy: incidence and treatment of psychiatric complications. Epilepsia. 1998;39:478–86.

Bob P. Dissociation and neuroscience: history and new perspectives. Int J Neurosci. 2003a; 113:903–14.

Bob P, Susta M, Chladek J, Glaslova K, Palus M. Chaos in schizophrenia associations, reality or metaphor? Int J Psychophysiol. 2009;73:179–85.

Bob P, Susta M, Glaslova K, Boutros NN. Dissociative symptoms and interregional EEG cross-correlations in paranoid schizophrenia. Psychiatry Res. 2010a;177:37–40.

Bob P, Palus M, Susta M, Glaslova K. Sensitization, epileptic-like symptoms and local synchronization in patients with paranoid schizophrenia. Prog Neuropsychopharmacol Biol Psychiatry. 2010b;34:143–6.

Bob P, Susta M, Gregusova A, Jasova D, Mishara A, Raboch J. Traumatic stress, dissociation, and limbic irritability in patients with unipolar depression being treated with SSRIs. Psychol Rep. 2010d;107:685–96.

Bogdan R, Pizzagalli DA. Acute stress reduces reward responsiveness: implications for depression. Biol Psychiatry. 2006;60:1147–54.

Bunge SA, Ochsner KN, Reskond JE, Dover GH, Gabrieli JDE. Prefrontal regions involved in keeping information in and out of mind. Brain. 2001;124:2074–86.

Castner SA, Williams GV. From vice to virtue: insights from sensitization in the nonhuman primate. Prog Neuropsychopharmacol Biol Psychiatry. 2007;31:1572–92.

Chaplin J, Yepez R, Shorvon S. A quantitative approach to measuring the social effects of epilepsy. Neuroepidemiology. 1990;9:151–8.

Chatrian GE, Bergamini L, Dondey M, et al. A glossary of terms most commonly used by clinical electroencephalographers. Electroencephalogr Clin Neurophysiol. 1974;37:538–48.

Cole DA, Nolen-Hoeksema S, Girgus J, Paul G. Stress exposure and stress generation in child and adolescent depression: a latent trait-state-error approach to longitudinal analyses. J Abnorm Psychol. 2006;115:40–51.

Collip D, Myin-Germeys I, Van Os J. Does the concept of "sensitization" provide a plausible mechanism for the putative link between the environment and schizophrenia? Schizophr Bull. 2008;34:220–5.

Costa E, Davis JM, Dong E, Grayson DR, Guidotti A, Tremolizzo L, et al. A GABAergic cortical deficit dominates schizophrenia pathophysiology. Crit Rev Neurobiol. 2004;16:1–23.

Critchley HD. Electrodermal Responses: what happens in the brain. Neuroscientist. 2002;8: 132–42.

Dranovsky A, Hen R. Hippocampal neurogenesis: regulation by stress and antidepressants. Biol Psychiatry. 2006;59:1136–43.

Duman RS, Monteggia LM. A neurothrophic model for stress-related mood disorders. Biol Psychiatry. 2006;59:1116–27.

Eack SM, Prasad KMR, Montrose DM, Goradia DD, Dworakowski D, Miewald J, et al. An integrated psychobiological predictive model of emergent psychopathology among young relatives at risk for schizophrenia. Prog Neuropsychopharmacol Biol Psychiatry. 2008;32:1873–8.

Eadie MJ. Dreamy mental states in late 20th century neurology. J Clin Neurosci. 1998;5:157–60.

Elbert T, Ray WJ, Kowalik ZJ, Skinner JE, Graf KE, Birbaumer N. Chaos and Physiology: Deterministic chaos in excitable cell assemblies. Physiol Rev. 1994;74:1–47.

Faber J, Vladyka V. Epileptogenesis and psychosogenesis, antithesis or synthesis? Acta Univ Carol Med. 1987;34:245–312.

Feinberg I. Efference copy and corollary discharge: implications for thinking and its disorders. Schizophr Bull. 1978;4:636–40.

Feinberg I, Guazzelli M. Schizophrenia- a disorder of the corollary discharge systems that integrate the motor systems of thought with the sensory systems of consciousness. Br J Psychiatry. 1999;17:4196–204.

Ferguson SM, Rayport M, Gardner R, Kass W, Weiner H, Reiser MF. Similarities in mental content of psychotic states, spontaneous seizures, dreams, and responses to electrical brain stimulation in patients with temporal lobe epilepsy. Psychosom Med. 1969;31:479–98.

Fisher R, van Emde BW, Blume W, Elger C, Genton P, Lee P, et al. Epileptic seizures and epilepsy: definitions proposed by the International League Against Epilepsy (ILAE) and the International Bureau for Epilepsy (IBE). Epilepsia. 2005;46:470–2.

Ford JM, Gray M, Faustman WO, Heinks TH, Mathalon DH. Reduced gamma-band coherence to distorted feedback during speech when what you say is not what you hear. Int J Psychophysiol. 2005;57:143–50.

Ford JM, Gray M, Faustman WO, Roach BJ, Mathalon DH. Dissecting corollary discharge dysfunction in schizophrenia. Psychophysiology. 2007;44:522–9.

Ford JM, Mathalon DH, Heinks T, Kalba S, Faustman WO, Roth WT. Neurophysiological evidence of corollary discharge dysfunction in schizophrenia. Am J Psychiatry. 2001;158: 2069–71.

Fries P. A mechanism for cognitive dynamics: neuronal communication through neuronal coherence. Trends Cogn Sci. 2005;9:474–80.

Fujiwara H, Namiki C, Hirao K, Miyata J, Shimizu M, Fukuyama H, et al. Anterior and posterior cingulum abnormalities and their association with psychopathology in schizophrenia: a diffusion tensor imaging study. Schizophr Res. 2007;28:215–22.

Gabriel A. Lamotrigine adjunctive treatment in resistant unipolar depression: an open, descriptive study. Depress Anxiety. 2006;23:485–8.

Glenthoj BY. The brain dopaminergic system. pharmacological, behavioural and electrophysiological studies. Dan Med Bull. 1995;42:1–21.

Glenthoj BY, Hemmingsen R. Dopaminergic sensitization: implications for the pathogenesis of schizophrenia. Prog Neuropsychopharmacol Biol Psychiatry. 1997;21:23–46.

Gloor P. Experiential phenomena of temporal lobe epilepsy. Facts and hypotheses. Brain. 1990;113:1673–94.

Gloor P, Olivier A, Quesney LF. The role of the limbic system in experimental phenomena of temporal lobe epilepsy. Ann Neurol. 1982;12:129–44.

Goddard GV. Development of epileptic seizures through brain stimulation at low intensity. Nature. 1967;214:1020–1.

Goldberg TE, Weinberger DR. Thought disorder in schizophrenia: a reappraisal of older formulations and an overview of some recent studies. Cogn Neuropsychiatry. 2000;5:1–19.

Gonzalez-Burgos G, Lewis DA. GABA neurons and the mechanisms of network oscillations: implications for understanding cortical dysfunction in schizophrenia. Schizophr Bull. 2008;34:944–61.

Grossman AW, Churchill JD, McKinney BC, Kodish IM, Otte SL, Greenough WT. Experience effects on brain development: possible contributions to psychopathology. J Child Psychol Psychiatry. 2003;44:33–63.

Halgren E, Walter RD, Cherlow DG, Crandall PH. Mental phenomena evoked by electrical stimulation of the human hippocampal formation and amygdala. Brain. 1978;101:83–117.

Harkness KL, Bruce AE, Lumley MN. The role of childhood abuse and neglect in the sensitization to stressful life events in adolescent depression. J Abnorm Psychol. 2006;115:730–41.

Heath RG. Brain function in epilepsy: midbrain, medullary, and cerebellar interaction with the rostral forebrain. J Neurol Neurosurg Psychiatry. 1976;39:1037–51.

Heath RG. Common characteristics of epilepsy and schizophrenia: clinical observation and depth electrode studies. 1961. Epilepsy Behav. 1961/2005;6:633–45.

Heckers S, Konradi C. Hippocampal neurons in schizophrenia. J Neural Transm. 2002;109: 891–905.

Hines M, Swan C, Roberts RJ, Varney NR. Characteristics and mechanisms of epilepsy spectrum disorder: An explanatory model. Applied Neuropsychol. 1995;2:1–6.

Hogan RE, Kaiboriboon K. The "dreamy state": John Hughlings-Jackson's ideas of epilepsy and consciousness. Am J Psychiatry. 2003;160:1740–7.

Hogan RE, Kaiboriboon K. John Hughlings-Jackson's writings on the auditory aura and localization of the auditory cortex. Epilepsia. 2004;45:834–7.

Hotchkiss AP, Harvey PD. Effect of distraction on communication failures in schizophrenic patients. Clin Res Rep. 1990;147:513–5.

Ito Y, Teicher M, Gold C, Harper D, Magnus E, Gelbard H. Increased prevalence of electrophysiological abnormalities in children with psychological, physical and sexual abuse. J Neuropsychiatry Clin Neurosci. 1993;5:401–8.

Jackson JH. Selected writings of John Hughlins Jackson. In: Taylor J, editor. On epilepsy and epileptiform convulsions, vol. 1. London: Hodder and Stoughton; 1931.

Jacob TC, Moss SJ, Jurd R. GABAA receptor trafficking and its role in the dynamic modulation of neuronal inhibition. Nat Rev Neurosci. 2008;9:331–43.

Jensen FE, Baram TZ. Developmental seizures induced by common early-life insults: short- and long-term effects on seizure susceptibility. Ment Retard Dev Disabil Res Rev. 2000;6:253–7.

Jobe PC, Dailey JW, Wernicke JF. A noradrenergic and serotoninergic hypothesis of the linkage between epilepsy and affective disorders. Crit Rev Neurobiol. 1999;13:317–56.

Johannessen LC. Antiepileptic drugs in non-epilepsy disorders: relations between mechanisms of action and clinical efficacy. CNS Drugs. 2008;22:27–47.

Kanner AM. Psychosis of epilepsy: a neurologist's perspective. Epilepsy Behav. 2000;1:219–27.

Kanner AM. The behavioural aspects of epilepsy: an overview of controversial issues. Epilepsy Behav. 2001;2:8–12.

Kanner AM, Balabanov A. Depression and epilepsy. How closely related are they? Neurology. 2002;58:827–39.

Keller MB. Past, present, and future directions for defining optimal treatment outcome in depression: remission and beyond. JAMA. 2003;289:3152–60.

Klonowski W, Jernajczyk W, Niedzielska K, Rydz A, Stepien R. Quantitative measure of complexity of EEG signal dynamics. Acta Neurobiol Exp. 1999;59:315–21.

Korn H, Faure P. Is there chaos in the brain? II. Experimental evidence and related models. C R Biol. 2003;326:787–840.

Kraus JE. Sensitization phenomena in psychiatric illness: lessons from the kindling model. J Neuropsychiatry Clin Neurosci. 2000;12:328–43.

Krishnamoorthy ES, Trimble MR, Sander JWAS, Kanner AM. Forced normalization at the interface between epilepsy and psychiatry. Epilepsy Behav. 2002;3:303–8.

Landolt H. Some clinical electroencephalographical correlations in epileptic psychosis (twilight states). Electroencephalogr Clin Neurophysiol. 1953;5:121.

Lee KH, Williams LM, Breakspear M, Gordon E. Synchronous gamma activity: a review and contribution to an integrative neuroscience model of schizophrenia. Brain Res Rev. 2003;41: 57–78.

Lee U, Mashour GA, Kim S, Noh GJ, Choi BM. Propofol induction reduces the capacity for neural information integration: implications for the mechanism of consciousness and general anesthesia. Conscious Cogn. 2009a;18;56–64.

Lee U, Mashour GA, Kim S, Noh GJ, Choi BM. Propofol induction reduces the capacity for neural information integration: implications for the mechanism of consciousness and general anesthesia. Conscious Cogn. 2009b;18;56–64.

Leroy F, Pezard L, Nandrino JL, Beaune D. Dynamical quantification of schizophrenic speech. Psychiatry Res. 2005;133:159–71.

Loscher W, Jackel R, Czucvar J. Is amygdala kindling in rats a model for drug-resistant partial epilepsy? Exp Neurol. 1986;93:211–26.

Mace CJ, Trimble MR. Psychosis following temporal lobe epilepsy: a report of six cases. J Neurol Neurosurg Psychiatry. 1991;54:639–44.

Marsh L, Rao V. Psychiatric complications in patients with epilepsy: a review. Epilepsy Res. 2002;49:11–33.

Mashour GA. Consciousness unbound: toward a paradigm of general anesthesia. Anesthesiology. 2004;100:428–33.

Mashour GA. Toward a general theory of unconscious processes in psychoanalysis and anesthesiology. J Am Psychoanal Assoc. 2008;56:203–22.

McCarley RW, Niznikiewicz MA, Salisbury DF, Nestor PG, O'Donnell BF, Hirayasu Y, et al. Cognitive dysfunction in schizophrenia: unifying basic research and clinical aspects. Eur Arch Psychiatry Clin Neurosci. 1999;249 Suppl 4:69–82.

McGrath J. Ordering thoughts on thought disorder. Br J Psychiatry. 1991;158:307–16.
McNamara JO, Morrisett R, Nadler JV. Recent advances in understanding mechanisms of the kindling model. Adv Neurol. 1992;57:555–60.
Meares R. The contribution of Hughlings Jackson to an understanding of dissociation. Am J Psychiatry. 1999;156:1850–5.
Meduna L. Über experimentelle campherepilepsie. Archiv für Psychiatrie. 1934;102:333–9.
Meduna L. Versuche über die biologische Beeinflussung des Ablaufes der Schizophrenie. Campher- und Cardiazolkräpfe Zeitschrift für Neurologie und Psychiatrie. 1935;152:235–62.
Möhler H. GABAA receptor diversity and pharmacology. Cell Tissue Res. 2006;326:505–16.
Monroe RR. Limbic ictus and atypical psychoses. J Nerv Ment Dis. 1982;170:711–6.
Monroe SM, Harkness KL. Life stress, the "kindling" hypothesis, and the recurrence of depression: considerations from a life stress perspective. Psychol Rev. 2005;112:417–45.
Monroe SM, Slavich GM, Torres LD, Gotlib IH. Major life events and major chronic difficulties are differentially associated with history of major depressive episodes. J Abnorm Psychol. 2007;116:116–24.
Morishita S. Clonazepam as a therapeutic adjunct to improve the management of depression: a brief review. Hum Psychopharmacol. 2009;24:191–8.
Nadel L, Samsonovich A, Ryan L, Moscovitch M. Multiple trace theory of human memory: computational, neuroimaging, and neuropsychological results. Hippocampus. 2000;10:352–68.
Nestor PG, Akdag SJ, O'Donnell BF, Niznikiewicz M, Law S, Shenton ME, et al. Word recall in schizophrenia: A connectionist model. Am J Psychiatry. 1998;155:1685–90.
Nestor PG, Han SD, Niznikiewicz M, Salisbury D, Spencer K, Shenton ME, et al. Semantic disturbance in schizophrenia and its relationship to the cognitive neuroscience of attention. Biol. Psychol. 2001;57:23–46.
Neligan A, Shorvon SD. Frequency and prognosis of convulsive status epilepticus of different causes: a systematic review. Arch Neurol. 2010;67:931–40.
Olypher AV, Klement D, Fenton AA. Cognitive disorganization in hippocampus: a physiological model of the disorganization in psychosis. J Neurosci. 2006;26:158–68.
Pakalnis A, Drake ME, Kuruvilla J, Blake JK. Forced normalization. Arch Neurol. 1988;45:139.
Paus T. Primate anterior cingulate cortex: where motor control, drive and cognition interface. Nat Rev Neurosci. 2001;2:417–24.
Peled A. Multiple constraint organization in the brain: a theory for schizophrenia. Brain Res Bull. 1999;49:245–50.
Penfield W. The excitable cortex in conscious man. Liverpool: Liverpool University Press; 1967.
Penfield W, Perot P. The brain's record of auditory and visual experience: a final summary and discussion. Brain. 1963;86:595–696.
Phillips TJ, Roberts AJ, Lessov CN. Behavioral sensitization to ethanol: genetics and the effects of stress. Pharmacol Biochem Behav. 1997;57:487–93.
Post RM. Transduction of psychosocial stress into the neurobiology of recurrent affective disorder. Am J Psychiatry. 1992;149:999–1010.
Post RM, Weiss SRB. A speculative model of affective illness cyclicity based on patterns of drug tolerance observed in amygdala-kindled seizures. Mol Neurobiol. 1996;13:33–60.
Post RM, Weiss SRB, Smith M, Li H, McCann U. Kindling versus quenching. implications for the evolution and treatment of posttraumatic stress disorder. Ann N Y Acad Sci. 1997;821: 285–95.
Post RM, Weiss SR, Pert A. Implications of behavioral sensitization and kindling for stress-induced behavioral change. Adv Exp Med Biol. 1988;245:441–63.
Post RM, Weiss RB. Sensitization and kindling phenomena in mood, anxiety, and obsessive-compulsive disorders: the role of serotonergic mechanisms in illness progression. Biol Psychiatry. 1998;44:193–206.
Post RM, Weis SR, Smith MA. Sensitization and kindling. In: Friedman MJ, Charney DS, Deutch AY, editors. Neurobiological and clinical consequences of stress: from normal adaptation to posttraumatic stress disorder. Philadelphia: Lipincott-Raven; 1995.

Poulet JF, Hedwig B. New insights into corollary discharges mediated by identified neural pathways. Trends Neurosci. 2007;30:14–21.
Putnam FW. Traumatic stress and pathological dissociation. Ann N Y Acad Sci. 1995;771: 708–15.
Putnam F. Dissociation in children and adolescents. A developmental perspective. London, New York: The Guilford Press; 1997.
Riklan M, Cullinan T, Cooper IS. Tension reduction and alerting in man following chronic cerebellar stimulation for the relief of spasticity or intractable seizures. J Nerv Ment Dis. 1977;164: 176–81.
Riklan M, Kabat C, Cooper IS. Psychological effects of short term cerebellar stimulation in epilepsy. J Nerv Ment Dis. 1976;162:282–90.
Roberts RJ, Gorman LL, Lee GP, Hines ME, Richardson ED, Riggle TA, et al. The phenomenology of multiple partial seizure like symptoms without stereotyped spells: an epilepsy spectrum disorder? Epilepsy Res. 1992;13:167–77.
Ryan L, Nadel L, Keil K, Putnam K, Schnyer D, Trouard T, et al. Hippocampal complex and retrieval of recent and very remote autobiographical memories: evidence from functional magnetic resonance imaging in neurologically intact people. Hippocampus. 2001;11:707–14.
Sammer G. Working-memory load and dimensional complexity of the EEG. Int J Psychophysiol. 1996;24:173–82.
Schmahmann J, Sherman JC. The cerebellar cognitive affective syndrome. Brain. 1998;121:561–79.
Schmitz EB, Robertson MM, Trimble MR. Depression and schizophrenia in epilepsy: social and biological risk factors. Epilepsy Res. 1999;35:59–68.
Schutter DLJG, van Honk J. An electrophysiological link between the cerebellum, cognition and emotion: frontal theta EEG activity to single-pulse cerebellar TMS. Neuroimage. 2006;33: 1227–31.
Segal ZV, Williams JM, Teasdale JD, Gemar M. A cognitive science perspective on kindling and episode sensitization in recurrent affective disorder. Psychol Med. 1996;26:371–80.
Servit Z, Musil F. Prophylactic treatment of posttraumatic epilepsy: results of a long-term follow-up in Czechoslovakia. Epilepsia. 1981;22:315–20.
Silberman E, Post R, Nurenberger J, Theodore W, Boulenger J. Transient sensory, cognitive, and affective phenomena in affective illness: a comparison with complex partial epilepsy. Br J Psychiatry. 1985;146:81–9.
Singer W. Synchronization of cortical activity and its putative role in information procesing and learning. Annu Rev Physiol. 1993;55:349–74.
Singer W. Consciousness and the binding problem. Ann N Y Acad Sci. 2001;929:123–46.
Smith PF, Darlington CL. The development of psychosis in epilepsy: a reexamination of the kindling hypothesis. Behav Brain Res. 1996;75:59–66.
Spigelman I, Li Z, Banerjee PK, Mihalek RM, Homanics GE, Olsen RW. Behavior and physiology of mice lacking the GABAA-receptor delta subunit. Epilepsia. 2002;43:3–8.
Sporns O, Tononi G, Edelman GM. Connectivity and complexity: the relationship between neuroanatomy and brain dynamics. Neural Netw. 2000;13:909–22.
Sporns O, Tononi G, Edelman GM. Theoretical neuroanatomy and the connectivity of the cerebral cortex. Behav Brain Res. 2002;135:69–74.
Sramka M, Sedlak P, Nadvornik P. Observation of kindling phenomenon in treatment of pain by stimulation in thalamus. In: Sweet W, editor. Neurosurgical treatment in psychiatry. Baltimore: University Park Press; 1977. p. 651–4.
Stevens JR. Schizophrenia: reproductive hormones and the brain. Am J Psychiatry. 2002;159: 713–9.
Stevens JR. Epilepsy, schizophrenia and the extended amygdala. Ann N Y Acad Sci. 1999;156: 548–61.
Svetlak M, Bob P, Cernik M, Kukleta M. Electrodermal complexity during the stroop colour word test. Autonomic Neuroscience: Basic and Clinical. 2010;152:101–7.
Teicher MH, Glod CA, Surrey J, Swett C, Jr. Early childhood abuse and limbic system ratings in adult psychiatric outpatients. J Neuropsychiatry Clin Neurosci. 1993;5:301–6.

Teicher MH, Andersen SL, Polcari A, Anderson CM, Navalta CP, Kim DM. The neurobiological consequences of early stress and childhood maltreatment. Neurosci Biobehav Rev. 2003;27:33–44.

Teicher MH, Tomoda A, Andersen SE. Neurobiological consequences of early stress and childhood maltreatment: Are results from human and animal studies comparable? Ann NY Acad Sci. 2006a;1071:313–23.

Tiihonen J, Wahlbeck K, Kiviniemi V. The efficacy of lamotrigine in clozapine-resistant schizophrenia: a systematic review and meta-analysis. Schizophr Res. 2009;109:10–4.

Tirsch WS, Stude Ph, Scherb H, Keidel M. Temporal order of nonlinear dynamics in human brain. Brain Res Rev. 2004;45:79–95.

Tononi G. Consciousness, information integration, and the brain. Prog Brain Res. 2005;150: 109–26.

Tononi G, Edelman GM. Schizophrenia and the mechanisms of conscious integration. Brain Res Rev. 2000;31:391–400.

Trimble MR. Anticonvulsant-induced psychiatric disorders. The role of forced normalization. Drug Saf. 1996;15:159.

Vaitl D, Lipp O, Bauer U, Schüler G, Stark R, Zimmerman M, et al. Latent inhibition in schizophrenia: Pavlovian conditioning of autonomic responses. Schizophr Res. 2002;55:147–58.

van de Grind W. Physical, neural, and mental timing. Conscious Cogn. 2002;11:241–64.

van Drongelen W, Lee HC, Hereld M, Chen Z, Elsen FP, Stevens RL. Emergent epileptiform activity in neural networks with weak excitatory synapses. IEEE Trans Neural Syst Rehabil Eng. 2005;13:236–41.

Van Os J, Rutten BPF, Poulton R. Gene-environment interactions in schizophrenia: review of epidemiological findings and future directions. Schizophr Bull. 2008;34:1066–82.

Van Winkel R, Stefanis NC, Myin-Germeys I. Psychosocial stress and psychosis. A review of the neurobiological mechanisms and the evidence for gene-stress interaction. Schizophr Bull. 2008;34:1095–105.

Varela FJ, Lachaux JP, Rodriguez E, Martinerie J. The brainweb: phase synchronization and large-scale integration. Nat Rev Neurosci. 2001;2:229–39.

Varney NR, Garvey MJ, Cook BL, Campbell DA, Roberts RJ. Identification of treatment-resistant depressives who respond favorably to carbamazepine. Ann Clin Psychiatry. 1993;5:117–22.

Vigo DV, Baldessarini RJ. Anticonvulsants in the treatment of major depressive disorder: an overview. Harv Rev Psychiatry. 2009;17:231–41.

Vignal J, Maillard L, McGonigal A, Chauvel P. The dreamy state: hallucinations of autobiographic memory evoked by temporal lobe stimulations and seizures. Brain. 2007;130(1):88–99.

Weng G, Bhalla US, Iyengar R. Complexity in biological signaling systems. Science. 1999;284: 92–6.

Wolf P. Acute behavioural symptomatology at disappearence of epileptiform EEG abnormality. Paradoxical or 'forced' normalization. Adv Neurol. 1991;55:127–42.

Wolf P, Trimble MR. Biological antagonism and epileptic psychosis. Br J Psychiatry. 1985;146: 272–6.

Yuii K, Suzuki M, Kurachi M. Stress sensitization in schizophrenia. Ann N Y Acad Sci. 2007;1113: 276–90.

Chapter 4
The Binding Problem and the Dissociated Mind

This chapter describes some of the current findings and hypotheses that might help to explain how disturbed interactions producing patterns of temporal disorganization and decreased or increased functional connectivity may underlie specific perceptual and cognitive states related to the disintegration of consciousness. The phenomenon of disintegrated consciousness is historically linked to Janet's (Fig. 4.1) research of divided consciousness and dissociation, which may occur under hypnosis or in psychopathological states, and also to Bleuler's concept of splitting in schizophrenia.

Dissociation is a special form of consciousness in which events that would ordinarily be connected are divided from one another (Li and Spiegel 1992). According to DSM-III-R and DSM-IV dissociation is defined as "a disturbance or alteration in the normally integrative functions of identity, memory or consciousness." Another definition is based on the disintegration of consciousness being the inability to integrate some psychic content into the consciousness (Bernstein and Putnam 1986). Nemiah (1981) defined the basic features of dissociation as an alteration of identity related to disturbances of memory of an individual during a dissociative state. Changes in identity may appear as depersonalization or in serious cases as dissociative identity disorder (multiple personality disorder in older terminology). A typical change in memory is, for example, psychogenic amnesia, and other typical symptoms are changes in the interpretation of the external world, such as derealization or hallucinations (Spiegel and Cardena 1991). Dissociation also produces typical somatoform symptoms, such as alterations in the sensation of pain (analgesia, kinesthetic anesthesia), painful symptoms, perception alterations, motor inhibition or loss of motor control, gastrointestinal symptoms (Nijenhuis et al. 1996, 1998, 2004; Nijenhuis 2000), and dissociative seizures (Brown and Trimble 2000; Kuyk et al. 1999).

P. Bob, *Brain, Mind and Consciousness: Advances in Neuroscience Research*,
DOI 10.1007/978-1-4614-0436-1_4, © Springer Science+Business Media, LLC 2011

Fig. 4.1 Pierre Janet

4.1 Dissociation and Its History

According to recent findings, a level of conscious integration may change during certain conditions related to experimental cognitive manipulations, hypnosis or stressful experiences that can lead to dissociation of consciousness. In psychopathological research the term dissociation was proposed by Pierre Janet to explain the processes related to the splitting of consciousness due to traumatic events or during hypnosis.

This area of research was instigated by Theódule Ribot, who, in his clinical work, investigated patients with diseases of the memory. For a theoretical explanation of his clinical data, Ribot used the principle of dissolution proposed by Hughlings Jackson (Ellenberger 1970), which stated that the behavior of an individual due to dissolution is more automatic, with less voluntary control and less complexity than in a normal state, because a loss of higher nervous functions develops later, leading to a dysregulation and exaggeration of more primitive functions (Ellenberger 1970; Meares 1999).

Following Ribot's findings, Pierre Janet wrote about the concept of dissociation as being a specific defensive response leading to memory loss (Ellenberger 1970). Pierre Janet, in his work on psychological automatisms, defined the process of loosening associations using the word *désagrégation*, which was later synonymously used with the term "dissociation" and became known mainly through the works of William James and Morton Prince (Ellenberger 1970). According to Janet's description, dissociation means a deficit of the associated system that creates the secondary consciousness (Janet 1890; Ellenberger 1970; van der Hart and Friedman 1989; van der Kolk and van der Hart 1989). Following hypnotic experiments with his teacher Charcot, Janet found that people under hypnosis experienced exceptional states of divided consciousness, which, in some cases manifested as "different personalities" (Janet 1890; Ellenberger 1970). This line of investigation later appeared in the psychoanalytical works of Sigmund Freud and Joseph Breuer, who considered secondary consciousness in *Studies in Hysteria* (Breuer and Freud 1895). Breuer and Freud conceptualized pathological conditions observed in conversion phenomena as a consequence of repression, and according to them, "dissociated states" were elicited by repression of the energy of the libido. In the early days of psychoanalysis Freud began his project on scientific psychopathology with the purpose of finding brain mechanisms related to cognitive functions that constitute normal and abnormal mental processes (Ellenberger 1970; Rofe 2008). His collaboration with Joseph Breuer uncovered a new development in psychology and provided a new conceptual framework for understanding of the mind–body problem, in which mental and somatic factors were closely connected and understood as different aspects of unity (Ellenberger 1970; Breuer and Freud 1895; Briquet 1859; Mace 1992).

In the same context, Janet elaborated on the concept of dissociation in his work *Psychological Automatism* (Havens 1966; Janet 1890; van der Hart and Friedman 1989), where he sketched out his notion of psychic functions and structures. He dealt with psychological phenomena that are often observable in hysteria, hypnosis, and states of suggestion or possession, and found that during complete psychological automatism related to psychological regression (Janet 1890; Ellenberger 1970; van der Hart and Friedman 1989), consciousness is totally dominated by repeating past experiences, such as in somnambulism or hysterical crises. Janet also described partial automatism, in which only a part of consciousness is dominated, and suggested that in these states unconscious mechanisms related to traumatic experiences that repress conscious control and perception might play an important role, calling them "subconscious fixed ideas" (Janet 1890; Ellenberger 1970). In the recent literature, the fixed idea is defined as formation of new spheres of consciousness around memories of intensely arousing experiences with a high emotional charge that organize the cognitive, affective, and visceral elements of the traumatic experience, while simultaneously keeping them out of conscious awareness (van der Hart and Friedman 1989). The fixed ideas may emerge in many forms of psychopathological or somatoform symptoms, for example, paroxysms, which may be understood to be a representation of psychological trauma when a fixed idea is transformed into hallucinations and body movements (van der Hart and Friedman 1989). Fixed ideas are presented in the form of dreams and dissociative episodes (e.g., hysterical attacks) or during hypnosis as a secondary consciousness. A characteristic feature of these states is a lowering of the mental level (*abaissement du niveau mental*), which is manifested by increased dissociation and mental depression connected to the reduction of psychological tension.

4.2 Dissociation and Traumatic Stress

Although dissociation is a concept proposed for the description of psychopathological phenomena, recent findings show that dissociation is not only pathological, but also involves some adaptive functions. Dissociation also occurs within a continuum of dissociative symptoms in the normal population and only in their abnormal form are dissociative symptoms attributed to mental disorder (Putnam 1989). Putnam, in agreement with historical tradition in his studies of dissociative reaction, suggested that most dissociative disorders might be induced by traumatic events (Putnam 1989, 1997), and other recent studies indicate that dissociative disorders are mainly induced because of a traumatic event (Sar and Ross 2006; Brewin 2007; Bob 2008). On the other hand, some studies show that brain insult, injury or other organic brain disease can play a role in this process (Kihlstrom 2005; Spiegel 1997), and in ICD-10 the organic dissociation, induced by a variety of conditions affecting cerebral function (Good 1993), is also defined. The concept of organic dissociation is based on evidence that dissociative symptoms and disorders, including amnesia, fugue, depersonalization, multiple personality, automatisms, and certain furors, can be induced by a variety of medications, abuse of drugs, and medical illnesses or conditions affecting CNS functions. It is important to note that organic dissociation can be distinguished from intoxication, amnestic disorder, and delirium (Good 1993).

According to recent evidence the most significant traumas inducing dissociative disorders originate in childhood and are most frequently related to physical or sexual abuse, with the subsequent development of symptoms, often after many years (Teicher et al. 2003; 2006a). However, traumatic events after serious accidents or natural disasters may also cause dissociative symptoms or disorders (Spiegel and Cardena 1991; Putnam 1997). For example, Chu and Dill (1990) investigated dissociation by means of the Dissociative Experiences Scale (DES) in 98 female subjects and found significantly higher dissociation in patients who had been exposed to emotional or physical abuse. Coons et al. (1989), in their study of the prevalence of traumas in childhood and in the adult clinical population, found that 100% of patients with atypical dissociative disorders and 82% diagnosed with psychogenic amnesia had documented physical or verbal abuse or neglect in childhood. About half of patients also experienced significant trauma in adulthood.

In this context, there is evidence that exposure to a significant psychological stressor preserves or even enhances memory of the emotional aspects of an event, and simultaneously disrupts memory of nonemotional aspects of the same event (Payne et al. 2006). For example, Briere and Conte (1989) have documented that 59.6% of a group of 468 patients with a proven history of sexual abuse in childhood were not able to remember episodes of abuse from the past. Similar results were found in a group of 87 children aged 7–15 years who had been exposed to a traumatic event requiring hospitalization, indicating that specifically children who showed temporal disorganization, but not absence of emotion or dissociative amnesia, in narrative themes, were more likely to report concurrent subsyndromal post-traumatic stress disorder (PTSD) symptoms at 4–7 weeks post-trauma (Kenardy et al. 2007).

Well-documented evidence also shows that individuals who are victims of a trauma in many cases are unable to register pain (for example, during self-injury) or painful effects (Butler et al. 1996; Frankel 1996; Agargun et al. 1998; Ebrinc 2002; Trief 1996; Saxe et al. 2002; Russr et al. 1993; Orbach et al. 1997), and that patients with dissociative disorders frequently report amnesia for self-injury (Saxe et al. 2002; Putnam 1989).

In addition, there is evidence that not only the effects of physical abuse, sexual abuse, or the witnessing of domestic violence, but also parental verbal aggression as a specific form of abuse may cause dissociation. Teicher et al. (2006a) collected data from 554 subjects 18–22 years of age (68% female), who responded to advertisements. They used the Verbal Abuse Questionnaire to assess exposure to parental verbal aggression and measures of dissociation, symptoms of limbic irritability, depression, anxiety, and anger/hostility. They found that exposure to verbal abuse alone and the witnessing of domestic violence had moderately strong effects on dissociation ratings and that subjects exposed to both verbal abuse and domestic violence (but no other form of maltreatment) had Dissociative Experience Scale scores 4.5 times higher than those of the nonabused subjects, which implies that parental verbal aggression is a serious form of child abuse and maltreatment that influences dissociation, symptoms of limbic irritability, depression, and other psychiatric symptoms (Teicher et al. 2006a).

The close relationship among the symptoms of traumatic stress, dissociation, limbic irritability, and depression has also been documented in depressive patients (Bob et al. 2010d). The authors found that the increased presence of traumatic stress symptoms in 113 unipolar depressive patients treated with SSRIs was associated with significantly more severe symptoms of dissociation, depression, and limbic irritability in comparison to 86 healthy controls. The results indicated that traumatic stress symptoms in depressive patients had a significant association with symptoms of dissociation, depression, and limbic irritability in both groups. In this context, there is growing evidence that traumatic experiences constitute a very important factor in many psychiatric disorders and that dissociative symptomatology often occurs because of child abuse, especially in cases of chronic emotional, physical or sexual abuse (Spiegel and Cardena 1991; Putnam 1997; Teicher et al. 2003; Sar and Ross 2006).

4.3 Neurobiological Consequences of Dissociation and Traumatic Stress

Dissociation as a reaction to psychological stressors and traumas has various neurobiological consequences (Teicher et al. 2003, 2006a; Putnam 1997). A basic feature of this stress activation is corticotropin-releasing hormone (CRH), which regulates the peripheral activities of the HPA axis, the sympathetic nervous activity, and immune responses (Chrousos et al. 1995; Elenkov et al. 2000; Elenkov and Chrousos 2002; Bob et al. 2010c). The CRH directly stimulates IL-6, IL-1, IL-12, and substance P production, and also upregulates TNF-alpha (Lotz et al. 1988; Leu and Singh 1992; Elenkov and Chrousos 2002). On the other hand the stress activation of the immune system predominantly produces various cytokines, such as IL-6, IL-1, and TNF-alpha, that coordinate and integrate the brain–endocrine immune response, and stimulate the hypothalamus, amygdala, and pituitary, which in turn stimulate CRH secretion and activate both the HPA axis and the sympathetic nervous system (Besedovsky et al. 1986; Kovacs and Elenkov 1995; Elenkov and Chrousos 2002). Other basic features of the stress activation are influences of proinflammatory cytokines, such as IL-6, IL-1, and TNF-alpha, on deficits in neurogenesis and BDNF expression that may explain cognitive and memory deficits related to traumatic stress and dissociation (Schulte-Herbruggen et al. 2005; Maes et al. 2009).

Repeated stressors and re-experiencing of the traumatic event in childhood frequently cause the delayed effects of severe psychological trauma that lead to enhancement of the self-preservative catecholamine states related to anger, fear, meaninglessness, and a blunting of the emotional responses of the attachment behavior associated with dysfunction of the locus coeruleus, amygdala, and hippocampal systems (Henry 1993, 1997). The functional defects in the hippocampus lead to decreasing inhibitory control of the hippocampus on the HPA axis and cause a positive feed-forward cascade of glucocorticoid levels (Bao et al. 2008). Recent data indicate that the most serious disturbances of the HPA axis caused by traumatic events, such as childhood abuse or neglect, in the first years of life, often have a long-term impact on emotional, behavioral, cognitive, social, and physiological functions (Horowitz et al. 1979; Ito et al. 1998; Heim et al. 2000; Orr and Roth 2000; Teicher et al. 2003; Read et al. 2001).

Additionally, recent findings indicate that the right hemisphere is more vulnerable to traumatic influences than the left (Henry 1993, 1997). The reason for this is likely the increased right hemispheric connection with the limbic system in comparison with the left hemisphere. The right (more often nondominant) hemisphere is also better connected to the autonomic nervous system and plays a predominant role in the physiological and cognitive aspects of emotional processing. It is more specialized than the left hemisphere in neuroendocrine and autonomic activation, and in the secretion of the stress hormones, corticotrophin-releasing factors, and cortisol (Spence et al. 1996; Sullivan and Gratton, 1999a,b; Wittling and Pfluger 1990; Schore 2001, 2002). Evidence for this lateralization is provided by studies dealing with the relationship between conditioned fear response and amygdala function that show that this activation is right hemisphere-dominant (La Bar et al. 1998). Also, it has been reported that partial kindling of the right and not the left amygdala induces a long-lasting increase in anxiety-like behavior (Adamec 1997, 1999), and that kindling in the right amygdala induces increased production of the corticotrophin-releasing factors (Adamec and McKay 1993). Recent evidence also suggests that the right amygdala is more involved in the storage of fearful faces and in the expression of the emotionally influenced memory of aversive experiences than the left (Morris et al. 1999; Isenberg et al. 1999; Coleman-Mensches and McGaugh 1995; Schore 2002).

Lateralized regulation of stress responses and emotionality-related processes, indicating a close relationship between stress and the right brain mechanisms, also occurs at the level of the medial prefrontal cortex (mPFC) (Sullivan and Gratton 1999a,b, 2002). Prelimbic and infralimbic regions of the mPFC have an influence on visceral motor regions, autonomic functions, and emotional expression, and are important in the integration of neuroendocrine and autonomic activity into the behavioral states and cognitive processes (Sullivan and Gratton 1999a,b, 2002). These studies additionally suggest that although the right mPFC is necessary for a normal stress response and adaptation, excessive activity of this region is predominantly maladaptive.

Influences of stress and dissociation on information transfer across the corpus callosum have also been observed (Spitzer et al. 2004). This interhemispheric dysfunction may indicate a relationship between traumatic dissociation on a psychological level and the related "functional dissociation" of the hemispheres. This functional dissociation, according to the literature, may be a form of reversible blocking of information transfer across the corpus callosum (Bogen and Bogen 1969). This might explain why certain dissociative symptoms are similar to symptoms in patients with split brain as a consequence of the anatomical "dissociation" between hemispheres that occurs after surgery on the corpus callosum (Ahern et al. 1993; Galin 1974; Brende 1984; Bob 2003b; Bogen and Bogen 1969; Spitzer et al. 2004). The functional dissociation may be a defense mechanism that enables the healthy hemisphere to inhibit the negative impulses from the dysfunctional hemisphere, similar to psychological dissociation inhibiting a certain negative psychological impulse that does not fit into the current cognitive scheme. In this context, Nasrallah (1985) suggested that one of the vital components of interhemispheric integration might be the inhibitory influence of the verbally expressive hemispheric consciousness (predominantly the left) on thoughts, intentions, and feelings from the other hemisphere.

4.4 Dissociation as a Hypersynchronous Epileptiform State

Recent studies have found frequent and unusual EEG abnormalities in victims of child abuse (Putnam 1997; Ito et al. 1993; Teicher et al. 1993, 2003). Typical EEG abnormalities found in traumatized and dissociated patients often involve temporal or frontal slow wave activity and may also involve frontotemporal spikes or sharp waves predominantly in the left hemisphere (Putnam 1997; Teicher et al. 2003). Although dissociative disorders cannot be generally explained on the basis of neurological dysfunction, contemporary data provide evidence that temporal lobe seizure activity can produce dissociative syndrome, which is similar to that observed in functional cases (Spiegel 1991). The evidence is based on the relatively frequent occurrence of epileptic discharges or epileptiform abnormalities documented in dissociated patients who are victims of child abuse (Coons et al. 1989; Teicher et al. 1993, 2003; Ito et al. 1993; Bob 2003a), during dissociative states, such as depersonalization (Sierra and Berrios 1998), multiple personality disorder (Mesulam 1981; Coons et al. 1982; Benson et al. 1986; Spiegel 1991), dissociative disorders not otherwise specified, and dissociative seizures (so-called pseudoepilepsy) (Bowman and Coons 2000). On the other hand there is evidence of relatively frequent ictal as well as interictal dissociative symptoms in epileptics, including multiple personality disorder (Schenk and Bear 1981; Ahern et al. 1993; Hersch et al. 2002). These findings suggest that temporal lobe epileptic activity might be important in the generation of dissociative symptoms without neurological focal lesions (Spiegel 1991). On the other hand there is also evidence that dissociative symptoms frequently occur in patients with temporal lobe epilepsy (Spiegel 1991). In addition, recent data suggest that the prevalence of seizure disorders is much higher in multiple personality disorder patients (Mesulam 1981; Schenk and Bear 1981; Benson 1986; Perrine 1991; Putnam 1997). For example, Ahern et al. (1993) examined the relationship of "multiple personality disorder" in two patients with temporolimbic epilepsy with certain types of hemispheric interaction. Both patients exhibited different "personalities" in a characteristic temporal relationship to their seizures, and were considered to be surgical candidates for the intracarotid sodium amobarbital procedure. They have demonstrated outbursts of emotional behavior during inactivation of the left hemisphere and these "different personalities" were known to the patient's families to manifest in the post-ictal period. These observations suggest that the association of multiple personality and temporolimbic epilepsy might not be dependent only on seizure discharges, but might also be related to certain types of hemispheric interaction (Ahern et al. 1993). Other dissociation-like symptoms, such as depersonalization, fugues, amnesias, and autoscopy (seeing an externalized image of oneself), were also reported ictally and peri-ictally by seizure patients (Putnam 1997). On the other hand, epileptic activity frequently occurs in dissociative phenomena, such as religious experiences (Saver and Rabin 1997) or out of body experiences (Blanke et al. 2002).

The frequent occurrence of epileptic activity and epileptiform abnormalities in dissociative states and disorders suggests that processes related to sensitization and kindling might be crucially important to an understanding of dissociative processes (Goddard et al. 1969) that can potentially explain how epileptic-like phenomena might arise from repeated trauma (Post et al. 1995; Putnam 1997; Teicher et al. 2003). The kindling related to traumatic stress, similar to experimental kindling, is likely caused by a progressively increasing response of groups of neurons due to repetitive subthreshold stimulation that may later lead to epileptic activity. Repeated stressful events likely determine sensitization, leading to an increase in responsiveness to stress stimuli resulting in significantly increased vulnerability to stressors that have more lasting consequences with kindling-like progression (Post et al. 1995; Post and Weiss 1998; Kraus 2000). The kindling model of stress-related sensitization (Post et al. 1995) seems to be in agreement with suggestive evidence that stress might influence the significantly increased occurrence of EEG abnormalities that have been reported in considerably traumatized patients (Teicher et al. 1993, 2003; 2006a; Putnam 1997; Ito et al. 1993). The kindling mechanism caused by stress may involve typical inhibitory failure related to overloading of defensive mechanisms, such as denial or "repression," that have been conceptualized in order to understand dissociative states (Yates and Nasby 1993). This process leads to a similar lack of inhibition to that demonstrated in epilepsy; therefore, it may also cause similar pathological electrophysiological changes.

4.5 Dissociation: Its Basic Neurobiological Mechanisms and Consequences for Psychotherapy

Recent findings suggest that dissociative states might be understood to be consequences of disturbances in memory consolidation and could be explained in the framework of this process. According to these findings stress may influence the atypical consolidation of short-term memory into long-term memory (Nadel and Jacobs 1998; McGaugh 2000; Debiec and Altemus 2006; Debiec and LeDoux 2006) and cause dissociation of memory systems concerned with encoding emotion and context on psychological, physiological, and anatomical levels (LeDoux 1992; LeDoux 1993; LeDoux 1994; Nadel and Jacobs 1998; Phillips and LeDoux 1992; Bechara et al. 1995).

The process of memory formation is, on a molecular level, linked to protein synthesis, which requires involvement of transcription factor CREB, expression of the brain-derived neurotrophic factor (BDNF), and other molecular processes that participate in the global processes of network consolidation, mainly in the hippocampus, but also in other brain structures (Debiec and Altemus 2006; Nadel and Jacobs 1998; Debiec et al. 2006; Lee et al. 2004). Transcription factor CREB is a protein that binds to specific DNA sequences and controls transcription of genetic information from DNA to mRNA in processes that enable neuronal plasticity and long-term memory formation in the brain (Silva et al. 1998). CREB specifically influences expression of BDNF, which, as a polypeptide growth factor, influences differentiation and survival of neurons in the nervous system and plays an important role in regulating synaptic plasticity, connectivity, and memory storage in the CNS (Bath and Lee 2006; Bramham and Messaoudi 2005). Interesting current findings indicate that this relationship between BDNF and cognitive processes is also significantly influenced by stress (Savitz et al. 2006) and that chronic stress in particular has an inhibitory influence on BDNF synthesis, synaptic plasticity, transmission, and memory formation, especially in the hippocampus and neocortex (Binder and Scharfman 2004; Thomas and Davies 2005; Savitz et al. 2006).

Additionally, there is evidence that decreased BDNF expression may significantly influence reparative processes and the neurogenesis of hippocampal neurons (Bremner 2006; Bremner et al. 2008), with resulting structural abnormalities of the hippocampus and other brain regions (Bremner et al. 1995; Bremner 1999; De Bellis et al. 1999, 2001; Jelicic and Merckecbach 2004; Winter and Irle 2004; Teicher et al. 2006a; Bremner et al. 2007). Meta-analyses of these reported studies show that in PTSD patients hippocampal volumes are significantly smaller than in controls with and without trauma exposure, but trauma-exposed patients without PTSD have also shown significantly smaller bilateral hippocampal volumes in comparison to nonexposed controls (Kitayama et al. 2005; Karl et al. 2006). Data reported by Choi et al. (2009) suggest a direct relationship between structural brain abnormalities and dissociative and traumatic stress symptoms. They studied 16 unmedicated subjects (4 male/12 female subjects, mean age 21.9 ± 2.4 years) with a history of high-level exposure to parental verbal abuse, but no other form of maltreatment, and 16 healthy control subjects (5 male/11 female subjects, 21.0 ± 1.6 years). They found that ratings of dissociation, limbic irritability, and depression were inversely associated with the integrity of the white matter tract in the left hippocampus (measured by fractional anisotropy) and ratings of somatization and anxiety were inversely correlated with the integrity of the white matter in the left fornix. Parental verbal abuse score was negatively significantly correlated with white matter integrity in the left superior temporal gyrus (Spearman r = –0.701), in the left hippocampus (Spearman r = –0.801), and in the left fornix (Spearman r = –0.524). Together, these findings indicate that stress through the molecular level affects the structure and connectivity of the brain (Nadel and Jacobs 1998; Payne et al. 2006; Bremner 2006; Bremner et al. 2008).

According to recent evidence, the influence of stress on BDNF decreases and other molecular processes specifically influence the fixation of new information by the process of long-term potentiation in a way that affects the spatiotemporal fragmentation that determines dissociative states (Nadel and Jacobs 1998; Binder and Scharfman 2004; Thomas and Davies 2005; Savitz et al. 2006; Payne et al. 2006). Current findings show that an extremely negative emotional experience during the traumatic event or inescapable stress likely blocks the induction of long-term potentiation in the hippocampus and mPFC, and influences atypical memory consolidation and memory distortion, which is characterized by the consolidation process predominantly on an implicit (subliminal) level in the amygdala. This blocking of higher-order behavior mediated by the hippocampus and mPFC allows more automatic responses, depending on the subcortical structures, mainly the amygdala (Bob 2007; Debiec and Altemus 2006; Nadel and Jacobs 1998; Payne et al. 2006; Vermetten and Douglas 2004; Maroun and Richter-Levin 2003). These findings are also in agreement with neuroimaging data that suggest that characteristic changes in the perfusion of limbic brain structures, such as the amygdala and the hippocampus, might coincide with high arousal and/or anxiety during traumatic recall (Vermetten and Douglas 2004). These findings suggest that a particular role in the specific formation of dissociative states is played by the amygdala, which participates in the modulation of memory consolidation and also plays a specific role in the consolidation of the traumatic memory (Cahill 1997; Cahill and McGaugh 1998; Bob 2007; Payne et al. 2006).

A typical feature of traumatic memories is that they are not acceptable for conscious awareness because of the accompanying strong negative emotions (Bob 2007; Nadel and Jacobs 1998; Payne et al. 2006). High anxiety and arousal related to traumatic experience are thought to focus the attention acutely and this attentional shift may produce fragmented memories, personality fragmentation (Bob 2007; Vermetten and Douglas 2004), psychological automatisms, and a lowering of the mental level, as described by Janet (Bob 2003a; Bob 2008; Ellenberger 1970; Frankel 1996; Havens 1966; Putnam 1997; Van Der Hart et al. 2005).

These data also imply important consequences for psychotherapy that must enable memory reconsolidation in safe and nonthreatening conditions, which leads to the neurobiological reprocessing of memory traces. This reconditioning and reconsolidation is therefore possible only by re-experiencing the traumatic memory in a new and safe situation, for example, during a psychotherapeutic process, which enables integration of the dissociative state (Bob 2007). During this process implicitly consolidated traumatic memory in subcortical structures, mainly in the amygdala, is probably transformed from the automatic to the higher level of conscious experience by long-term potentiation in the higher-level structures of the CNS, such as the mPFC and hippocampus.

In this context, Putnam proposed that the treatment of dissociative disorders and PTSD is based primarily on psychotherapeutic, cognitive, and behavioral modification approaches focused on "detoxifying" traumatic memories (Putnam 1995). These approaches to the psychotherapeutic treatment of dissociative disorders are historically linked to the term "abreaction" (van der Hart and Brown 1992). Abreaction is defined as an emotional release or discharge after recalling a painful experience that has been repressed because it was consciously intolerable (American Psychiatric Association 1987). This definition embodies historical controversy between the school of dissociation proposed by Janet and Charcot (Janet 1890; Ellenberger 1970) and later studies by Joseph Breuer and Sigmund Freud, who for an understanding of the abreactive process utilized the concept of repression in their *Studies of Hysteria* (Breuer and Freud 1895; Ellenberger 1970). Later Jung, following Janet, suggested that integration of traumatic memory related to the dissociated complex is a key process in the treatment of traumatized patients (Jung 1907). In agreement with these clinical data, further research and clinical practice indicate that repeated abreaction without integration of the dissociated state often has a malignant effect without any improvement in the patient's state and leads to intensification of intrusive symptoms (Brown and Fromm 1986; Horowitz 1986; Ross 1989; Van Der Hart and Brown 1992; Putnam 1992; Braun 1986).

The clinical evidence as well as the neuroscience of memory implies that without memory reconsolidation traumatic memories cannot be processed in an integrated mode of consciousness. In the case of abreactive experience, the successive reconsolidation provides a process during which the revivificated traumatic memory gets re-stabilized in its re-integrated form. In this context, neuroscientific research into memory and the emotional processes during traumatic recall induced by the abreactive process strongly suggests that successful therapeutic work with a dissociative state helps the individual, both psychologically and physiologically, and that measurable physiology is related to these changes induced by the psychotherapeutic process. The neural process of reconsolidation in principle may represent the potential existence of a new integrated and adaptive level in the neurophysiological process, which is actualized, for example, during successful therapy. Memory reconsolidation, therefore, likely presents a process that enables successful transformation from dissociated, automatic, and implicitly consolidated traumatic memory, mainly in the amygdala, to a higher level of conscious experience in the higher-level structures, such as the mPFC and hippocampus. This view corresponds to Janet's definition of the dissociative state being an automatic process that does not fit into the current cognitive scheme, and, without successful reprocessing (or reconsolidation) remains dissociated during recall of the dissociative state because of the specific neural substrate of dissociated memories (Bob 2007).

Additionally, this memory reconsolidation is again related to BDNF synthesis and other molecular processes that influence modification of synaptic plasticity, transmission, and new memory formation, especially in the hippocampus and mPFC, including gene expression, which is required for hippocampus-dependent long-term memory formation (Debiec et al. 2002; Morrison et al. 2003; Mizuno and Giese 2005; Savitz et al. 2006). These findings indicate that learning and memory processes, encompassing a wide variety of environmental factors, may also influence development of synaptic connections through new gene expression, and that psychotherapy, which is a special learning process, may specifically influence and modify brain functions, metabolism in specific brain structures, and also genetic processes (Kandel 1998, 1999; Gabbard 2000, 2007). From this point of view, in the future, a new era of psychotherapy research and practice may develop specific modes of psychotherapy designed to target specific sites of brain functioning (Gabbard 2000).

4.6 Dissociative States and Information Disintegration

Recent evidence indicates that brain functions related to consciousness and attention require multiregional functional interaction and large-scale integration and binding of multiple neural assemblies (Varela et al. 2001; Fries. 2005). According to several findings the interacting neuronal assemblies represent basic functional units in brain information processing that may behave independently with a lower level of binding among the units, or that may be dynamically integrated into large subsets of neurons that behave coherently with synchronous activity (Seth et al. 2006; Edelman 2003; Elbert et al. 1994; Lutzenberger et al. 1995; Molle et al. 1997; Stam 2005). This number of complex interlinked and simultaneously active neural states is reflected in the neural complexity, which is mathematically defined as a number of independent variables that determine the behavior of the system.

According to recent findings, there is evidence that neural complexity can be assessed according to EEG records. In this context several studies reported the relationship between attentional functions and brain EEG complexity, reflecting information processing. For example, Lutzenberger et al. (1995) reported that increased EEG complexity indicates an increase in simultaneously activated neural assemblies. Further studies also reported that EEG complexity was substantially greater during imagery than during actual sensory stimulation (Birbaumer et al. 1993; Schupp et al. 1994; Molle et al. 1997) or experimental cognitive load (Bizas et al. 1999; Meyer-Lindenberg 1996; Micheloyannis et al. 1998, 2002; Molle et al. 1995; 1997; Stam et al. 1996; Tomberg 1999). Further studies also indicate that complexity is significantly less during full alertness than during drowsiness (Matousek et al. 1995), and, similarly, clear alertness during a state of meditation has also been shown to be associated with a decrease in EEG complexity (Aftanas and Golocheikine 2002). Consistent with these data, it has been reported that divergent creative thought is associated with greater EEG complexity, whereas convergent analytical thought was related to the lower level of complexity (Molle et al. 1996). In this context other studies also show that neural EEG complexity reflects the attentional mode related to the processing of cortical stimuli (Pritchard and Duke 1995; Molle et al. 1995, 1996, 1997). Taken together, these results suggest that attentional narrowing might decrease complexity and cause a reduction in neural competition in connection with an inhibition of neural assemblies irrelevant to task completion during selective attention (Lutzenberger et al. 1992).

Preliminary data and models suggest that EEG complexity could in principle reflect typical attentional changes related to dissociative states and that dissociation could be described as a kind of divided or parallel neural process where several information processors within the brain system have a greater degree of independence (Li and Spiegel 1992, Bob 2003a). From this point of view dissociated consciousness is related to increased independence among neural assemblies and greater EEG complexity. However, the opposite is also true, because these states of increased neural complexity and decreased connectivity are interrupted by time periods when the dissociated state is released into consciousness, which leads to narrowing attention, with decreased complexity and increased connectivity, and information integration. These transient periods relating to an actual experience of aversive events or during reliving of a dissociative state lead to a greater allocation of attention that may cause changes in the ordinary integrative functions of consciousness (Guralnik et al. 2000; Vermetten and Bremner 2004; Bob 2008). Also, high anxiety and arousal related to disturbing past experience intensely narrow the attention (Vermetten and Bremner 2004) and a greater degree of functional connectivity and activation in certain brain regions was found in clinical forms of dissociation or hypnosis in functional imaging studies (Faymonville et al. 2006; Cojan et al. 2009). Together, these data suggest that dissociation in usual states of consciousness might be related to increased complexity; on the contrary, extreme levels of attention relating to the reliving of dissociated experiences or during hypnosis are linked to greater connectivity, a lower level of complexity, and increased autonomic and emotional arousal.

In this context, it is possible to suppose that complexity might also be related to a number of independent clusters of mental associations that are related to specific emotional states and autonomic arousal (Jung 1907; Weingartner et al. 1977; Stern and Riegel 1970; Bob et al. 2009a, 2009b, 2010). These complex patterns connecting mental and physiological states in turn produce specific patterns of temporal organization or disorganization with increased or decreased functional connectivity, which may underlie specific perceptual, emotional, and cognitive states (Sporns et al. 2000, 2002).

An extreme form of dissociation presents with loss of consciousness, which may occur, for example, in dissociative disorders not otherwise specified (American Psychiatric Association 1994; Sadock and Sadock 2008). In the unconscious, dissociated state we might expect heightened levels of dysconnectivity and complexity, which is in agreement with current findings in anesthesiology, indicating that the anesthetic-mediated interruption of consciousness cannot be explained by local CNS changes, but presents as a consequence of an "unbinding" mechanism across various parts of the brain (Mashour 2004, 2006; Lee et al. 2009a, 2009b). These data suggest that a common neurophysiology might underlie the loss of consciousness in anesthesia or a diminished level of consciousness in dissociative states.

Several findings also suggest that these deficits of conscious integration related to dissociation and increased complexity might conceptually be in agreement with theoretical concepts of neural processing, known as parallel distributed processing (PDP), or connectionist models of memory (Yates and Nasby 1993; Mc Clelland et al. 1986). From this point of view, dissociation can be described as a kind of divided or parallel neural process where several systems may have some independence (Li and Spiegel 1992; Butler et al. 1996). In this context, dissociation in principle could be understood using the concept of brain complexity to describe a level of parallel distribution in neural processing (Butler et al. 1996; Li and Spiegel 1992; Mc Clelland et al. 1986; Yates and Nasby 1993).

References

Adamec R. Transmitter systems involved in neural plasticity underlying increased anxiety and defense-implications for understanding anxiety following traumatic stress. Neurosci Biobehav Rev. 1997;21:755–65.

Adamec RE. Evidence that limbic neural plasticity in the right hemisphere mediates partial kindling induced lasting increases in anxiety-like behavior: effects of low frequency stimulation (Quenching?) on long-term potentiation of amygdala efferents and behavior following kindling. Brain Res. 1999;839:133–52.

Adamec RE, McKay D. Amygdala kindling, anxiety, and corticotopin releasing factor (CRF). Physiol Behav. 1993;54:423–31.

Aftanas LI, Golocheikine SA. Non-linear dynamic complexity of the human EEG during meditation. Neurosci Lett. 2002;330:143–6.

Agargun MY, Tekeoglu I, Kara H, Adak B, Ercan M. Hypnotizability, pain threshold, and dissociative experiences. Biol Psychiatry. 1998;44:66–71.

Ahern GL, Herring AM, Tackenberg J, Seeger JF. The association of multiple personality and temporolimbic epilepsy: intracarotid amobarbital test observations. Arch Neurol. 1993;50: 1020–5.

American Psychiatric Association. DSM III-R, diagnostic and statistical manual of mental disorders. 4th ed. Washington DC: American Psychiatric Association; 1987.

American Psychiatric Association. DSM IV, diagnostic and statistical manual of mental disorders. 4th ed. Washington DC: American Psychiatric Association; 1994.

Bao AM, Meynen G, Swaab DF. The stress system in depression and neurodegeneration: focus on the human hypothalamus. Brain Res Rev. 2008;57:531–53.

Bath KG, Lee FS. Variant BDNF (Val66Met) impact on brain structure and function. Cogn Affect Behav Neurosci. 2006;6:79–85.

Bechara A, Tranel D, Damasio H, Adolphs R, Rockland C, Damasio AR. Double dissociation of conditioning and declarative knowledge relative to the amygdala and hippocampus in humans. Science. 1995;269:1115–8.

Benson DF. Interictal behaviour disorders in epilepsy. Psychiatr Clin North Am. 1986;9:283–92.

Benson F, Miller BL, Signer SF. Dual personality associated with epilepsy. Arch Neurol. 1986;43:471–4.

Bernstein EM, Putnam FW. Development, reliability, and validity of a dissociation scale. J Nerv Ment Dis. 1986;174:727–35.

Besedovsky H, Del Rey A, Sorkin E, Dinarello CA. Immunoregulatory feedback between interleukin-1 and glucocorticoid hormones. Science. 1986;233:652–4.

Binder DK, Scharfman HE. Brain-derived neurotrophic factor. Growth Factors. 2004;22:123–31.

Birbaumer N, Lutzenberger W, Elbert T. Imagery and brain processes. In: Birbaumer N, Ohman A, editors. The structure of emotion. Toronto: Hogrefe and Huber; 1993. p. 122–38.

Bizas E, Simos PG, Stam CJ, Arvanitis S, Terzakis D, Micheloyannis S. EEG correlates of cerebral engagement in reading tasks. Brain Topogr. 1999;12:99–105.

Blanke O, Ortigue S, Landis T, Seeck M. Neuropsychology: stimulating illusory own-body perceptions. Nature. 2002;419:269–70.

Bob P. Dissociation and neuroscience: history and new perspectives. Int J Neurosci. 2003a;113: 903–14.

Bob P. Subliminal processes dissociation and the 'I'. J Anal Psychol. 2003b;48:307–16.

Bob P. Hypnotic abreaction releases chaotic patterns of electrodermal activity during dissociation. Int J Clin Exp Hypn. 2007;55:435–56.

Bob P. Pain, dissociation and subliminal self-representations. Conscious Cogn. 2008;17:355–69.

Bob P, Susta M, Gregusova A, Jasova D. Dissociation, cognitive conflict and nonlinear patterns of heart rate dynamics in patients with unipolar depression. Prog Neuropsychopharmacol Biol Psychiatry. 2009a;33:141–5.

Bob P, Susta M, Chladek J, Glaslova K, Palus M. Chaos in schizophrenia associations, reality or metaphor? Int J Psychophysiol. 2009b;73:179–85.

Bob P, Susta M, Glaslova K, Boutros NN. Dissociative symptoms and interregional EEG cross-correlations in paranoid schizophrenia. Psychiatry Res. 2010a;177:37–40.

Bob P, Susta M, Gregusova A, Jasova D, Mishara A, Raboch J. Traumatic stress, dissociation, and limbic irritability in patients with unipolar depression being treated with SSRIs. Psychol Rep. 2010b;107:685–96.

Bob P, Raboch J, Maes M, Susta M, Pavlat J, Jasova D, et al. Depression, traumatic stress and interleukin-6. J Affect Disord. 2010c;120:231–4.

Bob P, Susta M, Gregusova A, Jasova D, Mishara A, Raboch J. Traumatic stress, dissociation, and limbic irritability in patients with unipolar depression being treated with SSRIs. Psychol Rep. 2010d;107:685–96.

Bogen JE, Bogen GM. The other side of the brain: III. The corpus callosum and creativity. Bull Los Angel Neuro Soc. 1969;34:191–220.

Bowman ES, Coons PM. The differential diagnosis of epilepsy, pseudoseizures, dissociative identity disorder, and dissociative disorder not otherwise specified. Bull Menninger Clin. 2000;64: 164–80.

Bramham CR, Messaoudi E. BDNF function in adult synaptic plasticity: the synaptic consolidation hypothesis. Prog Neurobiol. 2005;76:99–125.

Braun BG. Issues in the psychotherapy of multiple personality disorder. In: Braun BG, editor. Treatment of multiple personality disorder. Washington, DC: American Psychiatric Press; 1986. p. 1–28.

Bremner JD. Does stress damage the brain. Biol Psychiatry. 1999;45:797–805.

Bremner JD. Stress and brain atrophy. CNS Neurol Disord Drug Targets. 2006;5:503–12.

Bremner JD, Randall P, Scott TM, Bronen RA, Seibyl JP, Southwick SM, et al. MRI-based measurement of hippocampal volume in combat-related posttraumatic stress disorder. Am J Psychiatry. 1995;152:973–81.

Bremner JD, Elzinga B, Schmahl C, Vermetten E. Structural and functional plasticity of the human brain in posttraumatic stress disorder. Prog Brain Res. 2007;167:171–86.

Bremner JD, Elzinga B, Schmahl C, Vermetten E: Structural and functional plasticity of the human brain in posttraumatic stress disorder. Prog Brain Res. 2008;167:171–86.

Brende JO. The psychophysiologic manifestations of dissociation: Electrodermal responses in a multiple personality patient. Psychiatr Clin North Am. 1984;7:41–50.

Breuer J, Freud S. Studies in hysteria. New York: Basic Books; 1895.

Brewin CR. Autobiographical memory for trauma: update on four controversies. Memory. 2007;15:227–48.

Briere J, Conte J. Amnesia and adults molested as children: testing theories of repression. Paper presented at the 97th Annual convention of the American Psychological Association, New Orleans; 1989.

Briquet P. Traité clinique et therapeutique de l'hystérie. Paris: JB Baillière; 1859.

Brown D, Fromm E. Hypnotherapy and hypnoanalysis. Hillsdale, NJ: L. Erlbaum Associates; 1986.

Brown RJ, Trimble MR. Dissociative psychopathology, non-epileptic seizures, and neurology. J Neurol Neurosurg Psychiatry. 2000;69:285–9.

Butler LD, Duran REF, Jasiukaitis P, Koopman CH, Spiegel D. Hypnotizability and traumatic experience: a diathesis-stress model of dissociative symptomatology. Am J Psychiatry, Festschrift Suppl. 1996;153:42–62.

Cahill L. The neurobiology of emotionally influenced memory. Implications for understanding traumatic memory. Ann N Y Acad Sci. 1997;821:238–46.

Cahill L, McGaugh JL. Mechanisms of emotional arousal and lasting declarative memory. Trends Neurosci. 1998;21:294–9.

Choi J, Jeong B, Rohan ML, Polcari AM, Teicher MH. Preliminary evidence for white matter tract abnormalities in young adults exposed to parental verbal abuse. Biol Psychiatry. 2009;65: 227–34.

Chrousos GP, Epstein F, Flier J, Reichlin S, Pavlou S. The hypothalamic-pituitary-adrenal axis and immune-mediated inflammation. N Engl J Med. 1995;332:1351–62.

Chu J, Dill D. Dissociative symptoms in relation to chidhood sexual and physical abuse. Am J Psychiatry. 1990;147:887–92.

Cojan Y, Waber L, Schwartz S, Rossier L, Forster A, Vuilleumier P. The brain under self-control: modulation of inhibitory and monitoring cortical networks during hypnotic paralysis. Neuron. 2009;62:862–75.

Coleman-Mensches K, McGaugh JL. Differential involvement of the right and left amygdalae in expression of memory for aversively motivated training. Brain Res. 1995;670:75–81.

Coons PM, Milstein V, Marley C. EEG studies of two multiple personality and a control. Arch Gen Psychiatry. 1982;39:823–5.

Coons PM, Bowman ES, Pellow TA. Post-traumatic aspects of the treatment of victims of sexual abuse and incest. Psychiatr Clin North Am. 1989;12:325–7.

De Bellis MD, Keshavan MS, Clark DB, Casey BJ, Giedd JN, Boring AM, et al. Developmental traumatology part II: brain development. Biol Psychiatry. 1999;45:1271–84.

De Bellis MD, Hall J, Boring AM, Frustaci K, Moritz G. A pilot longitudinal study of hippocampal volumes in pediatric maltreatment-related posttraumatic stress disorder. Biol Psychiatry. 2001;50:305–9.

Debiec J, Altemus M. Toward a new treatment for traumatic memories. Cerebrum. 2006;2–11.

Debiec J, LeDoux JE. Noradrenergic signaling in the amygdala contributes to the reconsolidation of fear memory: treatment implications for PTSD. Ann N Y Acad Sci. 2006;1071:521–4.

Debiec J, LeDoux JE, Nader K. Cellular and systems reconsolidation in the hippocampus. Neuron. 2002;36:527–38.

Debiec J, Doyere V, Nader K, LeDoux JE. Directly reactivated, but not indirectly reactivated, memories undergo reconsolidation in the amygdala. Proc Natl Acad Sci USA. 2006;103(9):3428–33.

Ebrinc S. Hypnotizability, pain threshold and dissociative experiences. Eur Neuropsychopharmacol. 2002;12(Suppl):427–8.

Edelman GM. Naturalizing consciousness: a theoretical framework. Proc Natl Acad Sci USA. 2003;100:5520–4.

Elbert T, Ray WJ, Kowalik ZJ, Skinner JE, Graf KE, Birbaumer N. Chaos and physiology: deterministic chaos in excitable cell assemblies. Physiol Rev. 1994;74:1–47.

Elenkov IJ, Chrousos GP. Stress hormones, proinflammatory and antiinflammatory cytokines, and autoimmunity. Ann N Y Acad Sci. 2002;966:290–303.

Elenkov IJ, Wilder RL, Chrousos GP, Vizi ES. The sympathetic nerve - an integrative interface between two supersystems: the brain and the immune system. Pharmacol Rev. 2000;52: 595–638.

Ellenberger HF. The discovery of the unconscious: the history and evolution of dynamic psychiatry. New York: Basic; 1970.
Faymonville ME, Boly M, Laureys SJ. Functional neuroanatomy of the hypnotic state. J Physiol Paris. 2006;99:463–9.
Frankel FH. Dissociation: the clinical realities. Am J Psychiatry. 1996;153(Suppl):64–70.
Fries P. A mechanism for cognitive dynamics: neuronal communication through neuronal coherence. Trends Cogn Sci. 2005;9:474–80.
Gabbard GO. A neurobiologically informed perspective on psychotherapy. Br J Psychiatry. 2000;177:117–22.
Gabbard GO. Psychotherapy in psychiatry. Int Rev Psychiatry. 2007;19:5–12.
Galin D. Implications for psychiatry of left and right cerebral specialization. Arch Gen Psychiatry. 1974;31:572–83.
Goddard GV, McIntyre GC, Leech CK. A permanent change in brain function resulting from daily electrical stimulation. Exp Neurol. 1969;25:295–330.
Good MI. The concept of an organic dissociative syndrome: what is the evidence? Harv Rev Psychiatry. 1993;1:145–57.
Guralnik O, Schmeidler J, Simeon D. Feeling unreal: cognitive processes in depersonalization. Am J Psychiatry. 2000;157:103–9.
Havens LL. Pierre Janet. J Nerv Ment Dis. 1966;143:383–98.
Heim C, Newport DJ, Heit S, Graham YP, Wilcox M, Bonsall R, et al. Pituitary-adrenal and autonomic responses to stress in women after sexual and physical abuse in childhood. JAMA. 2000;284:592–7.
Henry JP. Psychological and physiological responses to stress: the right hemisphere and the hypothalamo-pituitary-adrenal axis, an inquiry into problems of human bonding. Integr Physiol Behav Sci. 1993;28:369–87.
Henry JP. Psychological and physiological responses to stress: the right hemisphere and the hypothalamo-pituitary-adrenal axis, an inquiry into problems of human bonding. Acta Physiol Scand Suppl. 1997;640:10–25.
Hersch J, Yiu-Chung C, Smeltzer D. Identity shifts in temporal lobe epilepsy. Gen Hosp Psychiatry. 2002;24:185–7.
Horowitz MJ. Stress response syndromes. 2nd ed. Northvale, NJ: Jason Aronson; 1986.
Horowitz M, Wilner M, Alvarez W. Impact of event scale: a measure of subjective stress. Psychosom Med. 1979;41:209–18.
Isenberg N, Silbersweig D, Engelien A, Emmerich S, Malavade K, Beattie B, et al. Linguistic threat activates the human amygdala. Proc Natl Acad Sci USA. 1999;96:10456–9.
Ito Y, Teicher M, Gold C, Harper D, Magnus E, Gelbard H. Increased prevalence of electrophysiological abnormalities in children with psychological, physical and sexual abuse. J Neuropsychiatry Clin Neurosci. 1993;5:401–8.
Ito Y, Teicher MH, Glod CA, Ackerman E. Preliminary evidence for aberrant cortical development in abused children: a quantitative EEG study. J Neuropsychiatry Clin Neurosci. 1998;10: 298–307.
Janet P. L'Automatisme Psychologique. Paris: Felix Alcan; 1890.
Jelicic M, Merckecbach H. Traumatic stress, brain changes, and memory deficits: a critical note. J Nerv Ment Dis. 2004;192:548–53.
Jung CG. On psychophysical relations of the associative experiment. J Abnorm Psychol. 1907;1:247–55.
Kandel ER. A new intellectual framework for psychiatry. Am J Psychiatry. 1998;155:457–69.
Kandel ER. Biology and the future of psychoanalysis: a new intellectual framework for psychiatry revisited. Am J Psychiatry. 1999;156:505–24.
Karl A, Schaefer M, Malta LS, Dörfel D, Rohleder N, Werner A. A meta-analysis of structural brain abnormalities in PTSD. Neurosci Biobehav Rev. 2006;30:1004–31.
Kenardy J, Smith A, Spence SH, Lilley PR, Newcombe P, Dob R, et al. Dissociation in children's trauma narratives: an exploratory investigation. J Anxiety Disord. 2007;21:456–66.
Kihlstrom JF. Dissociative disorders. Annu Rev Clin Psychol. 2005;1:227–53.

Kitayama N, Vaccarino V, Kutner M, Weiss P, Bremner JD. Magnetic resonance imaging (MRI) measurement of hippocampal volume in posttraumatic stress disorder: a meta-analysis. J Affect Disord. 2005;88:79–86.
Kovacs KJ, Elenkov IJ. Differential dependence of ACTH secretion induced by various cytokines on the integrity of the paraventricular nucleus. J Neuroendocrinol. 1995;7:15–23.
Kraus JE. Sensitization phenomena in psychiatric illness: lessons from the kindling model. J Neuropsychiatry Clin Neurosci. 2000;12:328–43.
Kuyk J, Spinhoven P, Van Emde Boas W, Van Dyck R. Dissociation in temporal lobe epilepsy and pseudo-epileptic seizure patients. J Nerv Ment Dis. 1999;187:713–20.
La Bar KS, Gatenby JC, Gore JC, Le Doux JE, Phelps EA. Human amygdala activation during conditioned fear acquisition and extinction: a mixed-trial fMRI study. Neuron. 1998;20: 937–45.
LeDoux JE. Brain mechanisms of emotion and emotional learning. Curr Opin Neurobiol. 1992;2: 191–8.
LeDoux JE. Emotional memory systems in the brain. Behav Brain Res. 1993;58:69–79.
LeDoux JE. Emotion, memory and the brain. Sci Am. 1994;270:50–7.
Lee JL, Everitt BJ, Thomas KL. Independent cellular processes for hippocampal memory consolidation and reconsolidation. Science. 2004;304:839–43.
Lee U, Kim S, Noh GJ, Choi BM, Hwang E, Mashour GA. The directionality and functional organization of frontoparietal connectivity during consciousness and anesthesia in humans. Conscious Cogn. 2009a;18:1069–78.
Lee U, Mashour GA, Kim S, Noh GJ, Choi BM. Propofol induction reduces the capacity for neural information integration: implications for the mechanism of consciousness and general anesthesia. Conscious Cogn. 2009b;18:56–64.
Leu SC, Singh VK. Stimulation of interleukin-6 production by corticotropin-releasing factor. Cell Immunol. 1992;143:220–7.
Li D, Spiegel D. A neural network model of dissociative disorders. Psychiatr Ann. 1992;22: 144–47.
Lotz M, Vaughan JH, Carson DA. Effect of neuropeptides on production of inflammatory cytokines by human monocytes. Science. 1988;241:1218–21.
Lutzenberger W, Birbaumer N, Flor H, Rockstroh B, Elbert T. Dimensional analysis of the human EEG and inteligence. Neurosci Lett. 1992;143:10–4.
Lutzenberger W, Preissl H, Pulvermuller F. Fractal dimension of electroencephalographic time series and underlying brain processes. Biol Cybern. 1995;73:477–82.
Mace CJ. Hysterical conversion I: a history. Br J Psychiatry. 1992;161:369–77.
Maes M, Yirmyia R, Noraberg J, Brene S, Hibbeln J, Perini G, et al. The inflammatory and neurodegenerative (IandND) hypothesis of depression: leads for future research and new drug developments in depression. Metab Brain Dis. 2009;24:27–53.
Maroun M, Richter-Levin G. Exposure to acute stress blocks the induction of long-term potentiation of the amygdala-prefrontal cortex pathway in vivo. J Neurosci. 2003;23:4406–09.
Mashour GA. Consciousness unbound: toward a paradigm of general anesthesia. Anesthesiology. 2004;100:428–33.
Mashour GA. Integrating the science of consciousness and anesthesia. Anesth Analg. 2006;103: 975–82.
Matousek M, Wackermann J, Palus M, Berankova A, Albrecht V, Dvorak I. Global dimensional complexity of the EEG in healthy volunteers. Neuropsychobiology. 1995;31:47–52.
Mc Clelland JL, Rumelhart DE, PDP research group. Parallel distributed processing I,II. Cambridge: MIT Press; 1986.
McGaugh JL. Memory- a century of consolidation. Science. 2000;287:248–51.
Meares R. The contribution of Hughlings Jackson to an understanding of dissociation. Am J Psychiatry. 1999;156:1850–5.
Mesulam MM. Dissociative states with abnormal temporal lobe EEG. Arch Neurol. 1981;38: 176–81.
Meyer-Lindenberg A. The evolution of complexity in human brain development: an EEG study. Electroencephalogr Clin Neurophysiol. 1996;99:405–11.

Micheloyannis S, Flitzanis N, Papanikolau E, Bourkas M, Terzakis D, Arvanitis S, et al. Usefulness of non-linear EEG analysis. Acta Neurol Scand. 1998;97:13–9.
Micheloyannis S, Papanikolaou E, Bizas E, Stam CJ, Simos PG. Ongoing electroencephalographic signal study of simple arithmetic using linear and non-linear measures. Int J Psychophysiol 2002;44:231–38.
Mizuno K, Giese KP. Hippocampus-dependent memory formation: do memory type-specific mechanisms exist? J Pharmacol Sci. 2005;98:191–7.
Molle M, Marshall L, Pietrowsky R. Dimensional complexity of the EEG indicates a right fronto-cortical locus of attentional control. J Psychophysiol. 1995;9:45–55.
Molle M, Marshall L, Pietrowsky R. Enhanced dynamic complexity in the human EEG during creative thinking. Neurosci Lett. 1996;208:1–4.
Molle M, Albrecht C, Mashall L. Adrenocorticotropin widens the focus of attention in humans. A nonlinear electroencephalographic analysis. Psychosom Med. 1997;59:497–502.
Morris JS, Ohman A, Dolan RJ. A subcortical pathway to the right amygdala mediating "unseen" fear. Proc Natl Acad Sci USA. 1999;96:1680–5.
Morrison AP, Frame L, Larkin W. Relationships between trauma and psychosis: a review and integration. Br J Clin Psychol. 2003;42:331–53.
Nadel L, Jacobs WJ. Traumatic memory is special. Curr Dir Psychol Sci. 1998;7:154–7.
Nasrallah HA. The unintegrated right cerebral hemispheric consciousness as alien intruder: a possible mechanism for schneiderian delusions in schizophrenia. Compr Psychiatry. 1985;26: 273–82.
Nemiah JC. Dissociative disorders. In: Freeman AM, Kaplan HI, editors. Comprehensive textbook of psychiatry. 3rd ed. Baltimore: Williams and Wilkins; 1981.
Nijenhuis ERS. Somatoform dissociation: major symptoms of dissociative disorders. J Trauma Dissociation. 2000;1:7–32.
Nijenhuis ER, Spinhoven P, Van Dyck R, Van der Hart O, Vanderlinden J. The development and psychometric characteristics of the Somatoform Dissociation Questionnaire (SDQ-20). J Nerv Ment Dis. 1996;184:688–94.
Nijenhuis ERS, Spinhoven P, Vanderlinden J, van Dyck R, van der Hart O. Somatoform dissociative symptoms as related to animal defensive reactions to predatory imminence and injury. J Abnorm Psychol. 1998;107:63–73.
Nijenhuis ER, van der Hart O, Kruger K, Steele K. Somatoform dissociation, reported abuse and animal defence-like reactions. Aust N Z J Psychiatry. 2004;38:678–86.
Orbach I, Mikulincer M, King R, Cohen D, Stein D. Thresholds and tolerance of physical pain in suicidal and nonsuicidal adolescents. J Consult Clin Psychol. 1997;65:646–52.
Orr SP, Roth WT. Psychophysiological assessment: clinical applications for PTSD. J Affect Disord. 2000;61:225–40.
Payne JD, Jackson ED, Ryan L, Hoscheidt S, Jacobs JW, Nadel L. The impact of stress on neutral and emotional aspects of episodic memory. Memory. 2006;14:1–16.
Perrine KR. Psychopathology in epilepsy. Semin Neurol. 1991;11:175–81.
Phillips RG, LeDoux JE. Differential contribution of amygdala and hippocampus to cued and contextual fear conditioning. Behav Neurosci. 1992;106:274–85.
Post RM, Weiss RB. Sensitization and kindling phenomena in mood, anxiety, and obsessive-compulsive disorders: the role of serotonergic mechanisms in illness progression. Biol Psychiatry. 1998;44:193–206.
Post RM, Weis SR, Smith MA. Sensitization and kindling. In: Friedman MJ, Charney DS, Deutch AY, editors. Neurobiological and clinical consequences of stress: from normal adaptation to posttraumatic stress disorder. Philadelphia: Lipincott-Raven; 1995.
Pritchard WS, Duke D. Measuring "chaos" in the brain: a tutorial review of EEG dimension estimation. Brain Cogn. 1995;27:353–97.
Putnam FW. Diagnosis and treatment multiple personality disorder. New York, London: The Guilford Press; 1989.
Putnam FW. Using hypnosis for therapeutic abreactions. Psychiatr Med. 1992;10:51–65.
Putnam FW. Traumatic stress and pathological dissociation. Ann N Y Acad Sci. 1995;771: 708–15.

Putnam F. Dissociation in children and adolescents. A developmental perspective. London, New York: The Guilford Press; 1997.

Read J, Perry BD, Moskowitz A, Connolly J. The contribution of early traumatic events to schizophrenia in some patients: a traumagenic neurodevelopmental model. Psychiatry. 2001;64: 319–45.

Rofe Y. Does repression exist? Memory, pathogenic, unconscious and clinical evidence. Rev Gen Psychol. 2008;12:63–85.

Ross CA. Multiple personality disorder: Diagnosis, clinical features, and treatment. New York: John Wiley & Sons; 1989.

Russr MJ, Shearin EN, Clarkin JE, Harrison K, Hull JW. Subtypes of self-injurious patients with borderline personality disorder. Am J Psychiatry. 1993;150:1869–77.

Sadock BJ, Sadock VA. Kaplan and Sadock's concise textbook of clinical psychiatry. 3rd ed. Philadelphia, PA: Lippincott Williams, and Wilkins; 2008.

Sar V, Ross C. Dissociative as a confounding factor in psychiatric research. Psychiatr Clin North Am. 2006;29:129–44.

Saver JL, Rabin J. The neural substrates of religious experience. J Neuropsychiatry Clin Neurosci. 1997;9:498–510.

Savitz J, Solms M, Ramesar R. The molecular genetics of cognition: dopamine, COMT and BDNF. Genes Brain Behav. 2006;5:311–28.

Saxe GN, Chawla N, van der Kolk B. Self-destructive behavior in patients with dissociative disorders. Suicide Life Threat Behav. 2002;32:313–9.

Schenk L, Bear D. Multiple personality and related dissociative phenomena in patients with temporal lobe epilepsy. Am J Psychiatry. 1981;138:1311–6.

Schore AN. The effects of early relational trauma on right brain development, affect regulation and infant mental health. IMHJ. 2001;22:201–69.

Schore AN. Dysregulation of the right brain: a fundamental mechanism of traumatic attachment and the psychopathogenesis of posttraumatic stress disorder. Aust N Z J Psychiatry. 2002;36: 9–30.

Schulte-Herbruggen O, Nassenstein C, Lommatzsch M, Quarcoo D, Renz H, Braun A. Tumor necrosis factor-alpha and interleukin-6 regulate secretion of brain-derived neurotrophic factor in human monocytes. J Neuroimmunol. 2005;160:204–9.

Schupp HT, Lutzenberger W, Birbaumer N. Neurophysiological differences between perception and imagery. Cogn Brain Res. 1994;2:77–86.

Seth AK, Izhikevich E, Reeke GN, Edelman GM. Theories and measures of consciousness: an extended framework. Proc Natl Acad Sci USA. 2006;103:10799–804.

Sierra M, Berrios GE. Depersonalization: neurobiological perspectives. Biol Psychiatry. 1998;44: 898–908.

Silva AJ, Kogan JH, Frankland PW, Kida S. CREB and memory. Annu Rev Neurosci. 1998;21: 127–48.

Spence S, Shapiro D, Zaidel E. The role of the right hemisphere in the physiological and cognitive components of emotional processing. Psychophysiology. 1996;33:112–22.

Spiegel D. Neurophysiological correlates of hypnosis and dissociation. J Neuropsychiatry Clin Neurosci. 1991;3:440–5.

Spiegel D. Trauma, dissociation, and memory. Ann N Y Acad Sci. 1997;821:225–37.

Spiegel D, Cardena E. Disintegrated experience: the dissociative disorders revisited. J Abnorm Psychol. 1991;100:366–76.

Spitzer C, Willert C, Grabe H, Rizos T, Moller B, Freyberger HJ. Dissociation, hemispheric asymmetry, and dysfunction of hemispheric interaction: a transcranial magnetic stimulation approach. J Neuropsychiatry Clin Neurosci. 2004;16:163–9.

Sporns O, Tononi G, Edelman GM. Connectivity and complexity: the relationship between neuroanatomy and brain dynamics. Neural Netw. 2000;13:909–22.

Sporns O, Tononi G, Edelman GM. Theoretical neuroanatomy and the connectivity of the cerebral cortex. Behav Brain Res. 2002;135:69–74.

Stam CJ. Nonlinear dynamical analysis of EEG and MEG: review of an emerging field. Clin Neurophysiol. 2005;116:2266–301.

Stam CJ, van Woerkom TC, Pritchard WS. Use of non-linear EEG measures to characterize EEG changes during mental activity. Electroencephalogr Clin Neurophysiol. 1996;99:214–24.

Stern EJ, Riegel KF. Comparisons of the restricted association of chronic schizophrenic and normal control subjects. J Abnorm Psychol. 1970;75:164–71.

Sullivan RM, Gratton A. Lateralized effects of medial prefrontal cortex lesions on neuroendocrine and autonomic stress responses in rats. J Neurosci. 1999a;19:2834–40.

Sullivan RM, Gratton A. Medial prefrontal cortex, laterality, and stress. J Neurosci. 1999b;19: 2834–40.

Sullivan RM, Gratton A. Prefrontal cortical regulation of hypothalamic–pituitary–adrenal function in the rat and implications for psychopathology: side matters. Psychoneuroendocrinology. 2002;27:99–114.

Teicher MH, Glod CA, Surrey J, Swett Jr C. Early childhood abuse and limbic system ratings in adult psychiatric outpatients. J Neuropsychiatry Clin Neurosci. 1993;5:301–6.

Teicher MH, Andersen SL, Polcari A, Anderson CM, Navalta CP, Kim DM. The neurobiological consequences of early stress and childhood maltreatment. Neurosci Biobehav Rev. 2003;27: 33–44.

Teicher MH, Tomoda A, Andersen SE. Neurobiological consequences of early stress and childhood maltreatment: are results from human and animal studies comparable? Ann N Y Acad Sci. 2006a;1071:313–23.

Thomas K, Davies A. Neurotrophins: a ticket to ride for BDNF. Curr Biol. 2005;15:262–4.

Tomberg C. Focal enhancement of chaotic strange attractor dimension in the left semantic (Wernicke) human cortex during reading without concomitant change in vigilance level. Neurosci Lett. 1999;263:177–80.

Trief PM. Childhood abuse and chronic pain. Observations from clinical practice. J Interpers Violence. 1996;11:599–608.

Van der Hart O, Brown P. Abreaction re-evaluated. Dissociation. 1992;5:127–40.

Van der Hart O, Friedman B. A reader's guide to Pierre Janet on dissociation: a neglected intellectual heritage. Dissociation. 1989;2:3–16.

van der Hart O, Nijenhuis ER, Steele K. Dissociation: an insufficiently recognized major feature of complex posttraumatic stress disorder. J Trauma Stress. 2005;18:413–23.

van der Kolk BA, van der Hart O. Pierre Janet and the breakdown of adaptation of adaptation in psychological trauma. Am J Psychiatry. 1989;146:1530–40.

Varela FJ, Lachaux JP, Rodriguez E, Martinerie J. The brainweb: phase synchronization and large-scale integration. Nat Rev Neurosci. 2001;2:229–39.

Vermetten E, Bremner JD. Functional brain imaging and the induction of traumatic recall: a cross-correlational review between neuroimaging and hypnosis. Int J Clin Exp Hypn. 2004;52: 218–312.

Vermetten E, Douglas BJ. Functional brain imaging and the induction of traumatic recall: a cross-correlational review between neuroimaging and hypnosis. Int J Clin Exp Hypn. 2004;52: 280–312.

Weingartner H, Miller H, Murphy D. Mood-state-dependent retrieval of verbal associations. J Abnorm Psychol. 1977;86:276–84.

Winter H, Irle E. Hippocampal volume in adult burn patients with and without posttraumatic stress disorder. Am J Psychiatry. 2004;161:2194–200.

Wittling W, Pfluger M. Neuroendocrine hemisphere asymmetries: salivary cortisol secretion during lateralized viewing of emotion-related and neutral films. Brain Cogn. 1990;14:243–65.

Yates JL, Nasby W. Dissociation, affect, and network models of memory: an integrative proposal. J Trauma Stress. 1993;6:305–26.

Chapter 5
The Unconscious Mind

This chapter describes recent evidence that there is a limit of consciousness that presents a barrier between conscious and unconscious processes. The barrier is likely specifically related to disturbances of integrative neural mechanisms that through distributed brain processing linked to attentional mechanisms and memory enable integrative conscious experience to be formed. According to recent findings a level of conscious integration may change during certain conditions related to experimental cognitive manipulations, hypnosis or stressful experiences, which can lead to dissociation of consciousness. In this context, dissociation of consciousness likely presents a deficit in the global distribution of information that may result in a heightened level of independent neural processes and complexity in the brain.

Historical findings on unconscious processes in the late nineteenth and early twentieth century instigated great scientific discoveries that led to a fundamentally new view of the human mind. These findings defined the existence of a threshold between consciousness and the unconscious processes through which the unconscious spontaneously influences conscious activity and behavior (Bob 2003; Ellenberger 1970). The threshold of consciousness represents a limit of conscious experience, but in principle it is possible to assume that unconscious ideas, "knowledge," perception, intention, and affects represent existing entities that can be made accessible to the human consciousness and expand conscious awareness (Fig. 5.1; Jung 1972). At the same time, the edge of conscious experience is a fundamental problem for the human ability to know and understand, because the unconscious mind is dissociated from consciousness, and this dissociability indicates a distinction between contents of consciousness that actually exist and potential experience presented as a secondary consciousness that may exist behind the threshold.

Jung (1972, p. 173–8) suggests that such dissociation has two distinct aspects. The first aspect of this mental disunity consists of a possibility that the secondary consciousness as a potential conscious event was originally conscious and became subliminal because it does not fit into the dominant cognitive scheme. This situation

P. Bob, *Brain, Mind and Consciousness: Advances in Neuroscience Research*,
DOI 10.1007/978-1-4614-0436-1_5, © Springer Science+Business Media, LLC 2011

Fig. 5.1 Carl G. Jung

may typically occur because of a conflict in which the human mind cannot interpret some outside stimuli. In this context the secondary consciousness represents potential conscious experience, which may potentially present a new view of the world and the ability to interpret it.

The second aspect of "secondary consciousness" is that it may represent a potentially conscious event that is subliminal and never has been conscious. However, the unconscious mental representations that did not reach the edge of consciousness may indirectly influence the conscious mind and through this influence there is a possibility that the event might become a part of conscious experience (Stross and Shevrin 1968, 1969; Bob 2003; Shevrin 2001; Kihlstrom 1987, 2004). This approach is in principle based on the basic belief that all events occurring in the universe are causally related, which consequently means that a more detailed understanding of perceptual information through an increase in attentional sensitivity to stimuli may enable events to be discovered that are actually presented as hidden parts of human experience. These ideas suggest that attentional sensitivity might play a specific role in the process of understanding the world that, through more detailed introspective experience and increased sensitivity of measurement, enables scientific description of processes in the external material world.

5.1 Subliminal Perception and Limits of Consciousness

Recent findings in the neuroscience of consciousness strongly suggest that a level of synchronization and binding between various parts of the brain to some extent reflects the accessibility of various mental contents in the consciousness (Baars 2002; Diaz and McCarthy 2007; Melloni et al. 2007). These findings suggest that the "threshold of consciousness" reflects levels of information transmission and integration between distributed brain areas. In this context studies in cognitive neuroscience defined explicit and implicit perception, i.e., explicit perception means perception immediately presented to the consciousness of a subject, while implicit perception is not accessible in the awareness of the subject and cannot be verified immediately in the response of the subject, but only indirectly by observation or measurement of the subject's behavior or physiological response (Bob 2003; Kihlstrom 1987; Kihlstrom 2004). Implicit perception is in principle defined by findings indicating that cerebral cortical activities occurring as a response to a somatosensory stimulus must proceed for about 500 ms in order to elicit the conscious sensation (Libet et al. 1979; Libet et al. 1964, 1967, 1991). In this context, several research studies have examined the role of perceptual awareness in the processing of emotional pictures that were performed using the short projection of an image (Pessoa 2005; Wiens 2006). An interesting example is an experiment performed in 1957 by James Vicary. During a movie presentation two verbal messages, "Eat Popcorn" and "Drink Coke," were projected for 1/3,000 of a second at 5-s intervals, which led to increased popcorn consumption of about 58% and Coca Cola consumption of about 18% (Wortman et al. 1992; Karremans et al. 2006). Further studies confirmed a relative degree of autonomy related to emotional processing and found that responses to emotional stimuli in the human amygdala occur in the absence of awareness (Pessoa 2005; Wiens 2006). For example, analysis of the event-related-potentials (ERP) using a P3 wave (also called P300 – a positive wave with latency of 300 ms, after presentation of a stimulus) indicates that the P3 wave reflects neurophysiological changes associated with the subliminal stimulus connected with an emotional conflict in emotionally disturbed patients (Wong et al. 1994, 2004). Taken together, these findings and those of other reported studies provide evidence of subliminal perception and information processing (Poetzl 1960; Marcel 1983; Reanault et al. 1989; Roediger 1990; Crick and Koch 1995; Brazdil et al. 2001; Bunge et al. 2001; Bob 2003; Liddell et al. 2004; Gawronski et al. 2006; Kanwisher 2001; Shevrin 2001; Kihlstrom 1987, 2004), and show that unconscious attentional orienting specifically mediated by the amygdala, the subcortical retinotectal pathway, the superior colliculus, and several other structures, likely play a key role in the underlying mechanisms of subliminal processing (Mulckhuyse and Theeuwes 2010).

Although in principle a threshold of consciousness presenting absolute subliminality at which all discriminative responses disappear may be taken into account (Wortman et al. 1992; Erdelyi 2004a, 2004b; Reingold 2004), it seems likely that various sensory stimuli that have importance for cognitive processes and adaptive behavioral responses may be influenced by various mechanisms of cognitive modulation and that subliminality may represent a relative phenomenon characterized by a sensitivity of discriminative responses (Kihlstrom 2004).

Experimental conditions that enable modulatory influences on discriminative processes or attentional filtering to be assessed have also been studied using various methods of hypnotic suggestion, and several data indicate that the threshold of consciousness may change during hypnosis. For example, Stross and Shevrin

(1962, 1968, 1969) found alterations of thought contents under hypnosis in the investigation of "freely evoked images" after the subliminal presentation. These and other findings suggest that hypnosis might lead to heightened access to subliminal stimuli and that thought organization during hypnosis shares some common elements with thought organization during dreaming (Bob 2004). This finding corresponds to similar data that were reported when subliminally presented images were found in dreams (Fischer 1954; Poetzl 1960).

Further research using various methods of hypnotic modulation related to pain experience (Bob 2008; Villemure and Bushnell 2002) has also shown that highly hypnotizable persons possess stronger attentional filtering abilities than those who are less hypnotizable and that these differences are reflected in underlying brain dynamics, such as the interplay between the cortical and subcortical structures (Bob 2008; Crawford 1994; Eccleston and Crombez 1999; Feldman 2004). Highly hypnotizable persons can better focus and sustain their attention as well as better ignore irrelevant stimuli from the environment (Crawford 1994). This clinical experience corresponds to findings that descending inhibitory pathways, parallel to ascending sensory systems, can modulate quite early responses related to sensory information. Together, these findings suggest that high hypnotizable individuals can better inhibit incoming sensory stimuli. This inhibition likely emerges owing to the influence of the frontal cortex, which regulates the limbic system in the active gating of incoming sensory stimuli (Bob 2008; Crawford 1994). Together these findings indicate that the threshold of consciousness in hypnosis may significantly change and is dependent on exogenous and endogenous attention, which influences the content of perceptual awareness.

5.2 Perceptual Consciousness

Recent research on the neural correlates of perceptual consciousness consists of extensive evidence from behavioral studies on normal subjects as well as neurological patients that perceptual information can be represented in the mind/brain without the subject's awareness of that information (Crick and Koch 1995; Kanwisher 2001). These findings suggest that awareness of perceptual information requires not only a strong representation of the contents of awareness, but access to that information by other parts of the mind/brain (Baars 1988, 2002). The limits on conscious access to perceptual information may not be immutable. For example, some pathological brain changes may disrupt neural pathways, and as a consequence, perceptual information represented in one neural structure is not accessed by other parts of the system (Baars 2002; Kanwisher 2001).

According to this evidence even a strong neural representation may not be sufficient for awareness and there is behavioral evidence that perceptual awareness involves not only activation of the relevant perceptual properties, but also further construction of an organized representation in which these sensory properties are attributed to their sources in external objects and events (Kanwisher 2001; Baars 1988).

Fig. 5.2 Necker cube

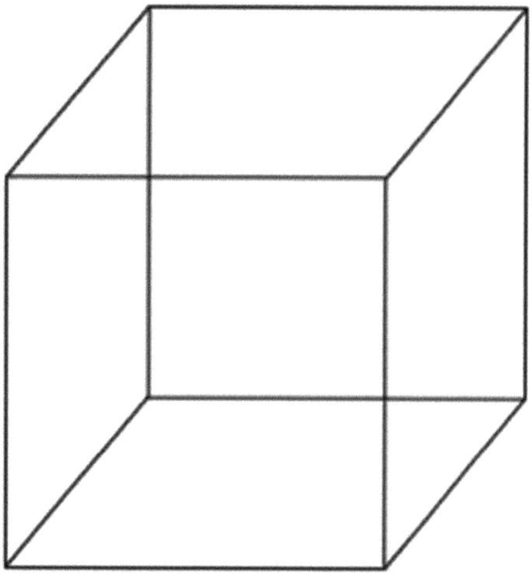

There is also evidence that only simple focusing of visual attention on different aspects of an unchanging stimulus has a strong effect on the content and intensity of perceptual awareness (Rees et al. 1999), and numerous studies using single-unit recordings, ERPs, and brain imaging have shown modulations of sensory responses by attention, even for a constant stimulus (Desimone and Duncan 1995; Luck and Girelli 1998; Goebel et al. 1998; Kanwisher 2001).

Simple examples of these cases provide ambiguous stimuli, leading to alterations in perceptual experience between two different states, such as the Necker cube (Fig. 5.2), Rubin's face/vase (Fig. 5.3), or experiments with binocular rivalry in which different images are projected for each eye (Kanwisher 2001; von Helmholtz 1962). Although the stimulus itself does not change, the human observer sees only one of the possible percepts, instead of seeing a mixture of the two possible images. For example, a typical experiment with binocular rivalry includes presentation of vertical stripes to the left eye and horizontal stripes to the right eye. The viewer is likely to see not a superimposition of the two patterns (i.e., a crosshatching, plaid pattern), but an alternating sequence in which only vertical stripes will be seen for one moment, and only horizontal stripes the next. Although the precise mechanisms underlying binocular rivalry are a matter of some debate, it is clear that experience alternates in a bistable fashion between being dominated by the input to one eye and being dominated by the input to the other eye (Blake et al. 1998; Kanwisher 2001).

Desimone and Duncan (1995) proposed a theoretical explanation of the observed phenomena in the concept of "interactive competition." According to this model, competitive interactions across cortical areas result in the domination of perceptual representations by the properties of a single object. This competition can be biased by either bottom–up factors (e.g., stimulus salience) or top–down factors (e.g., endogenous attention). In either case the result is that the various properties of an object represented in distinct cortical regions enhance each other and suppress

Fig. 5.3 Rubin's face/vase

the representation of competing objects with periods of dominance with mutual exclusivity (Rubin 2003), which results in multi-stability when the two (or more) percepts alternate in a seemingly random manner that likely has nonlinear dynamic features (Aks and Sprott 2003). In this context, attentional processing and awareness are global properties of the entire perceptual system that are linked to specific forms of binding, connecting multiple cortical areas, and these processes likely present a basic mechanism for perceptual consciousness, as well as for dissociative phenomena.

5.3 Attention and Dissociated Consciousness

Attention is a cognitive process of selective concentration on a particular stimulus while other competing stimuli are ignored and inhibited. Selective attention can be defined as a selection among potential conscious contents, and a specific function of attentional mechanisms is to bring different events to consciousness (Baars 1988, 2002). This process is related to the global distribution of information that is located in brain regions underlying conscious processing (Baars 1997, 1999, 2002). The event of attentional awareness is coupled with synchronized oscillatory activity in the gamma-band range that is related to Gestalt perception, and memory processing directly related to a stimulus (or endogenously generated) is also accompanied by increased synchrony between sensory and prefrontal regions (Fell et al. 2003; Kaiser and Lutzenberger 2003; Womelsdorf and Fries 2007; Sauseng and Klimesch 2008; Fries 2009).

In this context, there is evidence that consciousness may help to integrate brain functions that are separate and independent (Freeman 1991; Varela et al. 2001; Kanwisher 2001; Baars 2002). These findings also suggest that consciousness might be a gateway to brain integration that may enable access between otherwise separate neuronal functions, for example, in cases of executive dysfunctions in patients with dissociative identity disorder, dissociative fugue or during hypnosis (Baars 2002). In these cases a mutual dependence between the state of consciousness and executive input has been found and several memories were accessible in some states of consciousness while not in others, which suggests a loss of one executive interpreter's access to conscious events during the period of dominance of another interpreter (Hilgard 1986; Putnam et al. 1986; Putnam 1997; Baars 2002). Similar dissociation has been found in specific conscious states in split-brain patients, where each hemisphere's executive control over one side of the body, based on conscious input, may be limited to half of the visual field (Gazzaniga and Sperry 1967; Sperry 1968).

In agreement with these data, basic aspects of dissociation are related to selective specialization of cognitive functions, including a specific self-reference frame and identity, presenting a specific form of state-dependent memories resulting from a self-reflective experience. State-dependent memories are based on a principle that something learned in one neuropsychophysiological state is best recalled in the same state (Brown 1984). This phenomenon of state-dependent learning was studied in experiments with animals under the influence of drugs that influence memory formation in a neural state that it is not possible to recall without the drug. Similar experiments in humans confirm a relationship between a type of task and its recall (Brown 1984). These effects were also found in stress situations or in emotional states (Henry et al. 1973; Pearce et al. 1990), in sleep and circadian rhythmicity (Holloway 1978), post-ictal states (Overton 1978), as a consequence of changes in mood (Bower 1981), in cases of manic-depressive states induced by dextroamphetamine application (Henry et al. 1973), and using hypnosis (Brown 1984; Bower 1981).

Binding between the conscious contents and self-function observed in these cases constitutes the self-representational dimension of consciousness, which is characterized by interpretation of certain inner states of one's own body as a mental and somatic identity, while other bodily signals are interpreted as perceptions of the external world. Self-representations that are currently not accessible to the dominant interpreter's access are dissociated and may be defined as subliminal self-representations (Bob 2008). These subliminal self-representations as dissociated subsystems have sensory, emotional, and cognitive elements that may be misinterpreted and experienced as external reality, for example, during projection (or transference) when inner psychic states are interpreted as external parts of other persons, or during hallucinations, when certain internally generated voices or images are interpreted as sensory signals from the external world (Feinberg 1978; Feinberg and Guazzelli 1999; Ford et al. 2001; Tsakiris et al. 2005).

Together, these findings indicate that consciousness in dissociative states may significantly change and is dependent on exogenous and endogenous attention. Modulation of attention in dissociative states is coupled with global changes in subjective experience specifically focused on certain internal and external stimuli that may significantly influence the regulation and monitoring of the body, the mental state, experience of the self, and the underlying process of self-representation. These alterations in "self-representation" that underlie the changes in subjective experience are likely related to specific alterations in the binding of various parts of the brain, which defines the access of the dominant interpreter and the contents of awareness. Consequently, these alterations may be linked to great and abrupt changes in the patterns of neural activity that may cause dissociation of certain external and internal stimuli from awareness, and the lack of self-representation, which may lead to a distinct state of dissociated or divided consciousness (Crawford 1994; Rainville et al. 1997, 2002; Villemure and Bushnell 2002; Bob 2007; 2008; Vermetten and Douglas 2004).

5.4 Dissociative Identity Disorder and Basic Mechanisms of Dissociative States

Cognitive and affective representations of one's identity or the subject of experience present a basis for self-recognition as a specific cognitive process, typically involving conscious experience and interpretation activity. Disruptions of these self-interpretation processes likely represent a neurophysiological substrate for the process of the fragmentation of consciousness because of the misattribution of certain inner states that may be interpreted as external objects, because they are "disowned" and dissociated from consciousness. Psychological or physical stimuli leading to hopelessness and conflicting situations without known solutions that do not fit into the current cognitive scheme may lead to dissociation and "depersonalization" of certain perceptions, emotions, and cognitive strategies that create discrete "ego states" (or "alter personalities" in dissociative identity disorder), which are divided from the predominant state of consciousness (Bob 2008). All these processes of disrupted awareness and conscious integrity are likely related to and represented by similar disruptions at the brain level, and dissociation, in principle, may be explained by various levels of disturbed binding and brain coherence that may negatively affect usual patterns of connectivity, complexity, and synchronous activity constituting the adaptive integrative functions of consciousness (Bob 2010).

Dissociative identity disorder (in older terminology "multiple personality disorder"), characterized by two or more distinguished personalities, distinct identities or personality states existing in one person that recurrently take control of behavior and consciousness, is an interesting and controversial topic in the field of research on dissociation (Putnam 1989, 1997). At this point multiple personality disorder is a most complex form of dissociation related to a loss of the adaptive form of self-awareness and identity, suggesting that extremely stressful and traumatic conditions might cause switching within a spectrum of frames of self-reference. Dissociative identity disorder can be understood to be an extreme form of response to stress and social pressure that determines denial and amnesia of certain of one's own feelings, ideas, patterns of behavior, and resulting identity as a form of switching self-reference. These fluctuating states of consciousness and self-awareness within one human brain likely present a form of personality-state-dependent information processing typically manifested in the brain areas and networks involved in the experience of the self that has been reported by some neuroimaging data (Reinders et al. 2003).

These changes in self-perception, self-awareness, and self-representation are likely a consequence of profound changes in the affect state, memory, and sense of identity in response to environmental stress injury (Saxe et al. 2002), specifically related to changes in patterns of neural activity (Crawford 1994; Vermetten and Bremner 2004) and reflecting alterations in the binding of various parts of the brain related to self-representation and awareness (Bob 2008). These profound alterations also lead to significant changes in the interpretation of inner signals that have consequences for the sense of one's own identity and the external world. According to Helen Watkins' concept (Watkins 1993) these dissociated self-representations emerge as ego states (even in non-multiple subjects) and represent an organized cognitive structural system of segments of the personality that are often similar to true multiple personalities. The explanatory model is also based on reported data that these dissociated self-representations were reported in normal people under hypnosis and during dreaming in subjects with multiple personality disorder (Watkins and Watkins 1979–1980; Bowers and Brecher 1955; Gabel 1989; Watkins 1993; Lynn et al. 1994; Merskey 1992; Rickeport 1992; Barret 1995, 1996; Bob 2004).

The ego-state concept is in agreement with a series of intriguing case studies that strongly suggest the multiple personality structure to be a basic principle of the structural arrangement of the human mind, which in an extreme case may appear as dissociative identity disorder (Bob 2004). For example, Bowers and Brecher (1955) reported interesting material involved in the emergence of a multiple personality structure under hypnosis. The authors conclude that this structure was not produced by the hypnosis, but preceded the beginning of the hypnosis work. The patient in the case under discussion had not shown the multiple personality structure in clinical and psychological examinations prior to the hypnosis. In his conscious state the patient was not aware of his three underlying personalities, each of which reported distinctive dream material and Rorschach responses.

Similarly, Barret (1995) describes the relationship between the states of dreaming and multiple personality disorder, including amnesia and other alterations of memory, suggesting the dream character to be a hallucinated projection of aspects of the self that can be seen as a prototype for the alter personalities. Extreme early trauma may mutate or overdevelop these dissociated parts, inducing them to function in the external world, thus leading to development of the multiple personality disorder. According to these data the dream model parallels the observed phenomena of multiple personality disorder more directly than do explanations relying on waking fantasy processes (Barret 1995). In other studies, Barret (1994, 1996) reported cases of multiple personality with alters appearing as dream characters, or alters who could orchestrate dream content, and even cases of integration occurring within a dream, suggesting that the cognitive and personality processes operating outside conscious awareness might occur during dreaming.

Further literature also shows clinical evidence that the "dream work" of the ego is operative in both the representation of a separate self in dreams and in alter personalities (Brenner 1996, 1999, 2001). For example, a striking relationship between dreams and dissociative states was demonstrated in a patient with multiple personality disorder, who, in her usual state of consciousness reported a very distressing dream: she was watching a young girl being sexually abused by an unknown man while an unknown woman was holding her down. A number of days later, a young girl alter spontaneously emerged in a session, who described an eerily first-hand experience. This alter had no awareness that the dream had been reported and the patient had amnesia for the time when her alter was "out" giving her report of the trauma (Brenner 1996, 1999, 2001).

A very interesting case illustrating connections between dreams and the dissociated structure of the personality was reported by Salley (1988), whose patient, Frank, had multiple personality disorder, and found that his alter personalities emerged on parallel levels, on the one hand in dreams and on the other in hypnosis. Frank's multiple organization began at the age of 6 years, when he lived with his mother who was remarried to an alcoholic who abused Frank physically and emotionally. In his anamnesis was a wide spectrum of dissociative symptoms, such as a history of blackouts, amnesia for certain experiences, fugues, abrupt personality changes, and hysterical conversions. Frank had a long history of appearing in hospital grounds in a state of seizure with no memory of his identity, situation or past. Memory typically returned a few days after the seizure. During hypnotherapy, suggestive methods aimed at uncovering lost memory were used. While in a hypnotic trance Frank's subpersonality appeared, who identified himself as Self, a protector of Frank. Self in a somnambulistic trance explained that the seizures resulted from a struggle between Frank and Self. Self stated that his only line of communication with Frank was through dreams and that he would create a dream that would explain to Frank the function of the seizures. Out of the trance, Frank was as normal and had no memory of what had occurred under hypnosis. That night Frank dreamt that he was standing on a pedestal and two voices were shouting at him: one voice shouting "Yes!" and the other "No!" The vibrations from the shouting were so intense that the pedestal began to shake and split open, whereupon he fell to the ground shaking. Free association with the elements of the dream led Frank to relate the shaking to his seizures and the screaming to his internal conflict. In the 2 years since he had this dream, he has experienced no recurrence of the dissociative seizures, which suggests that the dream reflected communication between the dissociated and conflicting parts of his mind (Salley 1988).

Salley's case study and other documented dream works with patients suffering from multiple personality disorder and other trauma provide important data for research into the parallel levels between the dream and hypnotic states, suggesting that individual alter personalities might create dreams that provide therapeutically meaningful information, enable communication among dissociated parts of the personality, and provide access to the underlying personality structure (Ferenczi 1934; Levitan 1980; Putnam 1989; Jeans 1976; Marmer 1980a, 1980b; Hartmann 1998; Guralnik et al. 1999; Bob 2004).

Other interesting case studies describing multiple personality organization have been reported in chronic pain patients with dissociative identity disorder (Frances and Spiegel 1987; Livengood et al. 1994; McFadden and Woitalla 1993; McFadden 1992; Fishbain et al. 2001; Bob 2008). For example, a 34-year-old woman with chronic pain of the right wrist who developed dissociative identity disorder (Frances and Spiegel 1987) or a man who demonstrated another personality during the period of a severe, untreated headache (Livengood et al. 1994).

Conversely, it has been reported that the most common somatic complaint in patients with dissociative identity disorder is headache (Greaves 1980; Bliss 1980;

Coons et al. 1988; Coons et al. 1988; Coons 1980; Putnam et al. 1986) and symptoms of headaches in dissociative identity disorder patients have been described as extremely painful during the switching of personalities (Coons et al. 1988; Coons 1980; Larmore et al. 1977; Packard and Brown 1986). Several reported cases also suggest that patients with dissociative identity disorder can eliminate pain in the primary personality by displacing it into other alters (Watkins and Watkins 1990). Serious symptoms of headaches have also been reported in association with sudden unexpected travel away with amnesia and confusion about personal identity in patients with dissociative fugue (O'Brien 1985; Fishbain et al. 2001).

Together, these reports suggest that dissociative identity disorder might be considered an extreme case that might enable the basic structural model of the dissociated human personality to be found. This could in principle also explain other reported states of dissociation occurring in more usual clinical cases or, under exceptional experiences, in normal healthy individuals. Similar to the way in which we study structural determination of various forms in scientific disciplines like physics or chemistry, it might be possible to find the basic structural processes related to dissociation in mental processing. For such a description in principle, the basic structural components need to be found that enable the specific space for their interactions to be defined and the basic principles of these interactions to be formulated. This will enable us to understand why certain interacting states might create coherent wholes as a basis for integrative experience, or lead to disintegration, which determines discontinuous shifts among mental states.

The concept of discrete behavioral states, proposed by Frank Putnam (1997), provides a suitable formulation for an understanding of dissociative states, including the extreme cases of dissociation that occur in dissociative identity disorder. In this concept, behavioral states are described as the essential components of conscious experience involving a specific and unique configuration of a set of psychological, physiological, and behavioral states. This configuration creates psychoneurophysiological patterns of activation determining discrete patterns of thinking, feeling, and behavioral actions that regularly fluctuate in every human individual and have typical discontinuities in dissociative states.

The empirical background and definition of the discrete behavioral states originate from the study of infant mental states, which can be defined by a set of observable continuous and dichotomous behavioral states. Healthy children are born with a basic set of behavioral states and the number of infant states and their levels of interconnection increase with development. The growing behavioral repertoire leads to typical behavioral states in adulthood. This repertoire of behavioral states enables state space, which represents a set of typical behavioral patterns and states of mind, to be defined. It means that individual behavioral states exist within a larger multidimensional framework or space, defined by a set of states in the state space. An individual's behavior traverses the state space in a series of discontinuous jumps or switches from one state to another. The discrete states are transitory behavioral structures linked together by directional pathways forming behavioral architecture that defines an individual's personality. Transition between behavioral states manifests as a "switch" that represents an abrupt change in the values of the constellation of state-defining variables, for example, the transition from waking to sleeping, or in bipolar illness, from mania to depression. The model of discrete behavioral states defines "pathological dissociation" (Putnam 1995; Maaranen et al. 2005) as trauma-induced discrete behavioral states that are widely separated in multidimensional state space from normal states of consciousness in agreement with the conventional definition, which emphasizes the separation or segregation of specific ideas or affects from normal mental phenomena (Putnam 1997; Kaplan and Sadock 1991).

Putnam, in agreement with clinical experience, also proposed that observable differences between two discrete states are not a simple function of moving up or down and that these processes could be related to nonlinear dynamic features (Putnam 1997). In this context, for example, Wolff (1987) highlights differential responsiveness as an example of the nonlinearity of the input–output relation in different states of consciousness, and conceptualizes the relevance of the nonlinear dynamic systems theory to discrete behavioral states where switches between behavioral states constitute nonlinear transitions. Further recent studies provide stronger evidence that rapid shifts in mood and behavior could be related to nonlinear dynamic processes (Putnam 1997; Gottschalk et al. 1995).

In summary, the concept of discrete behavioral states defines structural components that enable changes in the mental space to be described based on their interactions, and provides descriptions of these interactions as continuous or discontinuous switching from one state to another. Based on these interactions the concept in principle enables interactions among ego states or subpersonalities in dissociative identity disorder to be qualitatively described, and generally such description enables the concept of order or disorder among behavioral states to be considered. These levels of order or disorder may in principle be understood within the concept of complexity, describing the level of independency among discrete behavioral states. This level of independency and complexity increases in dissociative disorders in comparison to a normal healthy state of mind, because of an antagonism among certain interacting behavioral states that cannot create coherent wholes as a basis for integrative experience, which leads to discontinuous shifts among mental states in certain conditions that are normally experienced as coherent and integrated. Consequently, these alterations in "self-representation," underlying specific ego-state-related changes in subjective experience, are likely related to specific alterations in the binding of various parts of the brain and these alterations may be linked to great and abrupt changes in patterns of neural activity. Using this concept, it is possible to imagine the abstract space of mind, represented by specific forms of coherent and antagonistic neural patterns. These patterns contain all the mental states experienced by an individual in a hierarchical order of specific types of mental experiences and their behavioral and neural patterns as objectively existing state-dependent potentialities that may be activated during waking states, dreaming, and other states of consciousness. Using this model, it might in principle be possible to consider various levels of order and disorder in the human mind and the brain using physical concepts of nonlinear dynamics, describing levels of entropy and complexity, and also to describe specific relationships between the mind and brain using physical and mathematical methods.

References

Aks DJ, Sprott JC. Resolving perceptual ambiguity in the Necker cube: a dynamical systems approach. J Nonlinear Dynamics Psychol Life Sci. 2003;7:159–78.

Baars BJ. A cognitive theory of consciousness. Cambridge: Cambridge University Press; 1988.

Baars BJ. Some essential differences between consciousness and attention, perception, and working memory. Conscious Cogn. 1997;6:363–71.

Baars BJ. Attention versus consciousness in the visual brain: differences in conception, phenomenology, behaviour, neuroanatomy, and physiology. J Gen Psychol. 1999;12:224–33.

Baars BJ. The conscious access hypothesis: origins and recent evidence. Trends Cogn Sci. 2002;6:47–52.

Barret D. Dreams in dissociative disorders. Dreaming. 1994;4:165–75.

Barret D. The dream character as prototype for the multiple personality alter. Dissociation. 1995;8: 66–8.

Barret D. Dreams in multiple personality disorder. In: Barret D, editor. Trauma and dreams. Cambridge: Harvard University Press, Cambridge, Massachusetts; 1996. p. 68–81.

Blake R, Yu K, Lokey M, Norman H. Binocular rivalry and motion perception. J Cogn Neurosci. 1998;10:46–60.

Bliss EL. Multiple personalities. A report of 14 cases with implications for schizophrenia and hysteria. Arch Gen Psychiatry. 1980;37:1388–97.

Bob P. Subliminal processes dissociation and the 'I'. J Anal Psychol. 2003;48:307–16.

Bob P. Dissociative processes, multiple personality and dream functions. Am J Psychother. 2004;58:139–49.

Bob P. Hypnotic abreaction releases chaotic patterns of electrodermal activity during dissociation. Int J Clin Exp Hypn. 2007;55:435–56.

Bob P. Pain, dissociation and subliminal self-representations. Conscious Cogn. 2008;17:355–69.

Bob P, Susta M, Glaslova K, Boutros NN. Dissociative symptoms and interregional EEG cross-correlations in paranoid schizophrenia. Psychiatry Res. 2010;177:37–40.

Bower GH. Mood and memory. Am Psychol. 1981;36:129–48.

Bowers MK, Brecher S. The emergence of multiple personalities in the course of hypnotic investigation. Int J Clin Exp Hypn. 1955;3:188–99.

Brazdil M, Rektor I, Daniel P, Dufek M, Jurak P. Intracerebral event-related potentials to subthreshold target stimuli. Clin Neurophysiol. 2001;112:650–61.

Brenner I. The characterological basis of multiple personality. Am J Psychother. 1996;52: 154–66.

Brenner I. Deconstructing DID. Am J Psychother. 1999;53:344–60.

Brenner I. Dissociation of trauma: theory, phenomenology and technique. Boston: International Universities Press; 2001.

Brown BG. Towards a theory of multiple personality and other dissociative phenomena. Psychiatr Clin North Am. 1984;7:171–93.

Bunge SA, Ochsner KN, Reskond JE, Dover GH, Gabrieli JDE. Prefrontal regions involved in keeping information in and out of mind. Brain. 2001;124:2074–86.

Coons PM. Multiple personality: diagnostic considerations. J Clin Psychiatry. 1980;41:330–6.

Coons PM, Bowman ES, Milstein V. Multiple personality disorder. A clinical investigation of 50 cases. J Nerv Ment Dis. 1988;176:519–27.

Crawford HJ. Brain dynamics and hypnosis. Int J Clin Exp Hypn. 1994;42:204–32.

Crick F, Koch C. Are we aware of neural activity in primary visual cortex? Nature. 1995;375: 121–3.

Desimone R, Duncan J. Neural mechanisms of selective visual attention. Annu Rev Neurosci. 1995;18:193–222.

Diaz MT, McCarthy G. Unconscious word processing engages a distributed network of brain regions. J Cogn Neurosci. 2007;19:1768–75.

Eccleston C, Crombez G. Pain demands attention: a cognitive-affective model of the interruptive function of pain. Psychol Bull. 1999;125:356–66.
Ellenberger HF. The discovery of the unconscious: the history and evolution of dynamic psychiatry. New York: Basic; 1970.
Erdelyi MH. Comments on commentaries: Kihlstrom, Bachmann, Reingold, and Snodgrass. Conscious Cogn. 2004a;13:430–3.
Erdelyi MH. Subliminal perception and its cognates: theory, indeterminacy, and time. Conscious Cogn. 2004b;13:73–91.
Feinberg I. Efference copy and corollary discharge: implications for thinking and its disorders. Schizophr Bull. 1978;4:636–40.
Feinberg I, Guazzelli M. Schizophrenia- a disorder of the corollary discharge systems that integrate the motor systems of thought with the sensory systems of consciousness. Br J Psychiatry. 1999;17:4196–204.
Feldman JB. The neurobiology of pain, affect and hypnosis. Am J Clin Hypn. 2004;46:187–200.
Fell J, Fernández G, Klaver P, Elger CE, Fries P. Is synchronized neuronal gamma activity relevant for selective attention? Brain Res Rev. 2003;42:265–72.
Ferenczi S. Gedanken über das trauma. International Zeitschrift für Psychoanalyse. 1934;20:5–12.
Fischer C. Dreams and perception: the role of preconscious and primary modes of perception and dream formation. J Am Psychoanal Assoc. 1954;2:389–445.
Fishbain DA, Cutler RB, Rosomoff HL, Rosomoff RS. Pain-determined dissociation episodes. Pain Med. 2001;2:216–24.
Ford JM, Mathalon DH, Heinks T, Kalba S, Faustman WO, Roth WT. Neurophysiological evidence of corollary discharge dysfunction in schizophrenia. Am J Psychiatry. 2001;158: 2069–71.
Frances A, Spiegel D. Chronic pain masks depression, multiple personality disorder. Hospital & Community Psychiatry. 1987;38:933–35.
Freeman WJ. The physiology of perception. Sci Am. 1991;264:78–85.
Fries P. Neuronal gamma-band synchronization as a fundamental process in cortical computation. Annu Rev Neurosci. 2009;32:209–24.
Gabel S. Dreams as a possible reflection of dissociated self-monitoring system. J Nerv Ment Dis. 1989;177:560–8.
Gawronski B, Hofmann W, Wilbur CJ. Are "implicit" attitudes unconscious? Conscious Cogn. 2006;15:485–99.
Gazzaniga MS, Sperry RW. Language after section of the cerebral commissures. Brain. 1967;90:131–48.
Goebel R, Khorram-Sefat D, Muckli L, Hacker H, Singer W. The constructive nature of vision: direct evidence from functional magnetic resonance imaging studies of apparent motion and motion imagery. Eur J Neurosci. 1998;10:1563–73.
Gottschalk AM, Bauer MS, Whybrow PC. Evidence of chaotic mood variation in bipolar disorder. Arch Gen Psychiatry. 1995;52:947–59.
Greaves GB. Multiple personality. 165 years after Mary Reynolds. J Nerv Ment Dis. 1980;168: 577–96.
Guralnik O, Levin R, Schmeidler J. Dreams of personality disordered subjects. J Nerv Ment Dis. 1999;187:40–6.
Hartmann E. Nightmare after trauma as paradigm for all dreams: a new approach to the nature and functions of dreaming. Psychiatry. 1998;61:223–38.
Henry G, Weingartner H, Murphy LD. Influence of affective states and psychoactive drugs on verbal learning and memory. Am J Psychiatry. 1973;130:66–71.
Hilgard ER. Divided consciousness. Multiple control in human thought and action. New York: Wiley; 1986.
Holloway FA. State dependent retrieval based on time of day. In Ho BT, Richards DV, Chute DL, editors. Drug discrimination and state dependent learning. Academic Press: New York; 1978. pp. 319–43.

Jeans RF. The three faces of Evelyn: a case report I. An independently validated case of multiple personalities. J Abnorm Psychol. 1976;85:249–55.
Jung CG. On the nature of the psyche. The structure and dynamics of the psyche. Collected Works of CG Jung. vol 8, Princeton: Princeton University Press; 1972.
Kaiser J, Lutzenberger W. Induced gamma-band activity and human brain function. Neuroscientist. 2003;9:475–84.
Kanwisher N. Neural events and perceptual awareness. Cognition. 2001;79:89–113.
Kaplan HI, Sadock BJ. Comprehensive glossary of psychiatry and psychology. Baltimore: Williams and Wilkins; 1991.
Karremans J, Stroebe W, Claus J. Beyond Vicary's fantasies: the impact of subliminal priming and brand choice. J Exp Soc Psychol. 2006;42:792–8.
Kihlstrom JF. The cognitive unconscious. Science. 1987;237:1445–52.
Kihlstrom JF. Availability, accessibility, and subliminal perception. Conscious Cogn. 2004;13: 92–100.
Larmore K, Ludwig AM, Cain RL. Multiple personality: an objective case study. Br J Psychiatry. 1977;31:3540.
Levitan H. The dream in traumatic states. In: Natterson JM, editor. The dream in clinical practice. New York: Jason Aronson; 1980.
Libet B, Alberts WW, Wright Jr EW, Delattre LD, Levin G, Feinstein B. Production of threshold levels of conscious sensation by electrical stimulation of human somatosensory cortex. J Neurophysiol. 1964;27:546–78.
Libet B, Alberts WW, Wright Jr EW, Feinstein B. Responses of human somatosensory cortex to stimuli below threshold for conscious sensation. Science. 1967;158:1597–600.
Libet B, Wright Jr EW, Einstein B, Pearl DK. Subjective refferal of the timing for a conscious sensory experience. Brain. 1979;102:193–224.
Libet B, Pearl DK, Morledge DA, Gleason CA, Hosobuchi Y, Barbaro NM. Control of the transition from sensory detection to sensory awareness in man by the duration of a thalamic stimulus: the cerebral 'time-on' factor. Brain. 1991;114:1731–57.
Liddell BJ, Williams LM, Rathjen J, Shevrin H, Gordon E. A temporal dissociation of subliminal versus supraliminal fear perception: an event-related potential study. J Cogn Neurosci. 2004;16: 479–86.
Livengood JM, Zouny CM, Parris CV. Implications of multiple personality disorder for treatment of chronic pain. Pain Digest. 1994;4:1914.
Luck SJ, Girelli M. Electrophysiological approaches to the study of selective attention in the human brain. In: Parasuraman R, editor. The attentive brain. Cambridge: MIT Press; 1998. p. 71–94.
Lynn SJ, Maré C, Kvaal S, Segal D, Sivec H. The hidden observer, hypnotic dreams, and age regression: clinical implications. Am J Clin Hypn. 1994;37:130–42.
Maaranen P, Tanskanen A, Honkalampi K, Haatainen K, Hintikka J, Viinamaki H. Factors associated with pathological dissociation in the general population. Aust N Z J Psychiatry. 2005;39: 387–94.
Marcel AJ. Conscious and unconscious perception: an approach to the relations between phenomenal experience and perceptual processes. Cogn Psychol. 1983;15:238–300.
Marmer SS. The dream in dissociative states. In: Natterson JM, editor. The dream in clinical practice. New York: Jason Aronson; 1980a.
Marmer SS. Psychoanalysis of multiple personality. Int J Psychoanal. 1980b;61:439–59.
McFadden IJ. Multiple personality disorder becoming manifest after persistence of a chronic pain syndrome. Clin J Pain. 1992;8:63.
McFadden IJ, Woitalla VF. Differing reports of pain perception by different personalities in a patient with chronic pain and multiple personality disorder. Pain. 1993;55:37982.
Melloni L, Molina C, Pena M, Torres D, Singer W, Rodriguez E. Synchronization of neural activity across cortical areas correlates with conscious perception. J Neurosci. 2007;27: 2858–65.
Merskey H. The manufacture of personalities. The production of multiple personality disorders. Br J Psychiatry. 1992;160:327–40.

Mulckhuyse M, Theeuwes J. Unconscious attentional orienting to exogenous cues: a review of the literature. Acta Psychol (Amst). 2010;134:299–309.
O'Brien M. The diagnosis of multiple personality syndromes: overt, covert, and latent. Compr Ther. 1985;11:59–66.
Overton DA. Major theories of state dependent learning. In: Ho BT, Richards DV, Chute DL, editors. Drug discrimination and state dependent learning. London: Academic Press; 1978.
Packard RC, Brown F. Multiple headaches in a case of multiple personality disorder. Headache. 1986;26:99102.
Pearce S, Isherwood S, Hrouda D, Richardson P, Erskine A, Skinner J. Memory and pain; tests of mood congruity and state dependent learning in experimentally induced and clinical pain. Pain. 1990;43:187–93.
Pessoa L. To what extent are emotional visual stimuli processed without attention and awareness? Curr Opin Neurobiol. 2005;15:188–96.
Poetzl O. The relationship between experimentally induced dream images and indirect vision. Psychological Issues Monograph. 1960;2:46–106.
Putnam FW. Diagnosis and treatment multiple personality disorder. New York, London: The Guilford Press; 1989.
Putnam FW. Traumatic stress and pathological dissociation. Ann N Y Acad Sci. 1995;771:708–15.
Putnam F. Dissociation in children and adolescents. A developmental perspective. London, New York: The Guilford Press; 1997.
Putnam FW, Guroff JJ, Silberman EK, Post BL. The clinical phenomenology of multiple personality disorder. Review of 100 recent cases. J Clin Psychiatry. 1986;47:285–9.
Rainville P, Duncan GH, Price DD, Carrier B, Bushnell MC. Pain affect encoded in human anterior cingulate but not somatosensory cortex. Science. 1997;277:968–71.
Rainville P, Hofbauer RK, Bushnell MC, Duncan GH, Price DD. Hypnosis modulates activity in brain structures involved in the regulation of consciousness. J Cogn Neurosci. 2002;14: 887–901.
Reanault B, Signoret JL, Debruille B, Breton F, Bolgert F. Brain potentials reveal covert facial recognition in prosopagnosia. Neuropsychologia. 1989;27:905–12.
Rees G, Russell C, Frith CD, Driver J. Inattentional blindness versus inattentional amnesia for fixated but ignored words. Science. 1999;286:2504–7.
Reinders AATS, Nijenhuis ERS, Paans AMJ, Korf J, Willemsen ATM, den Boer JA. One brain, two selves. Neuroimage. 2003;20:2119–25.
Reingold EM. Unconscious perception: assumptions and interpretive difficulties. Conscious Cogn. 2004;13:117–22.
Rickeport MM. The interface between multiple personality, spirit mediumship and hypnosis. Am J Clin Hypn. 1992;34:168–77.
Roediger HL. Implicit memory: retention without remembering. Am Psychol. 1990;45:1043–56.
Rubin N. Binocular rivalry and perceptual multi-stability. Trends Neurosci. 2003;26:289–91.
Salley RD. Subpersonalities with dreaming functions in a patient with multiple personalities. J Nerv Ment Dis. 1988;176:112–5.
Sauseng P, Klimesch W. What does phase information of oscillatory brain activity tell us about cognitive processes? Neurosci Biobehav Rev. 2008;32:1001–13.
Saxe GN, Chawla N, van der Kolk B. Self-destructive behavior in patients with dissociative disorders. Suicide Life Threat Behav. 2002;32:313–9.
Shevrin H. Event-related markers of unconscious processes. Int J Psychophysiol. 2001;42: 209–18.
Sperry RW. Hemisphere deconnection and unity in conscious awareness. Am Psychol. 1968;23: 723–33.
Stross L, Shevrin H. Differences in thought organization between hypnosis and the waking state: an experimental approach. Bull Meninger Clin. 1962;26:237–47.
Stross L, Shevrin H. Thought organization in hypnosis and the waking state. J Nerv Ment Dis. 1968;147:272–88.
Stross L, Shevrin H. Hypnosis as a method for investigating unconscious thought processes. J Am Psychoanal Assoc. 1969;17:100–35.

Tsakiris M, Haggard P, Franck N, Mainy N, Sirigu A. A specific role for efferent information in self-recognition. Cognition. 2005;96:215–31.

Varela FJ, Lachaux JP, Rodriguez E, Martinerie J. The brainweb: phase synchronization and large-scale integration. Nat Rev Neurosci. 2001;2:229–39.

Vermetten E, Bremner JD. Functional brain imaging and the induction of traumatic recall: a cross-correlational review between neuroimaging and hypnosis. Int J Clin Exp Hypn. 2004;52:218–312.

Vermetten E, Douglas BJ. Functional brain imaging and the induction of traumatic recall: a cross-correlational review between neuroimaging and hypnosis. Int J Clin Exp Hypn. 2004;52: 280–312.

Villemure Ch, Bushnell MC. Cognitive modulation of pain: how do attention and emotion influence pain processing? Pain. 2002;95:195–99.

von Helmholtz H. Treatise on physiological optics. New York: Dover; 1962.

Watkins HH. Ego-state therapy: an overview. Am J Clin Hypn. 1993;35:232–40.

Watkins JG, Watkins HH. Ego states and hidden observers. J Altered States of Consciousness. 1979–80;5:3–18.

Watkins JG, Watkins HH. Dissociation and displacement: where goes the "ouch?". Am J Clin Hypn. 1990;33:1–21.

Wiens S. Subliminal emotion perception in brain imaging: findings, issues, and recommendations. Prog Brain Res. 2006;156:105–21.

Wolff PH. The development of behavioural and emotional states in infancy. Chicago: University Chicago Press; 1987.

Womelsdorf T, Fries P. The role of neuronal synchronization in selective attention. Curr Opin Neurobiol. 2007;17:154–60.

Wong PS, Shevrin H, Williams WJ. Conscious and nonconscious processes: an ERP index of an anticipatory response in a conditioning paradigm using visual masked stimuli. Psychophysiology. 1994;31:87–101.

Wong PS, Bernat E, Snodgrass M, Shevrin H. Event-related brain correlates of associative learning without awareness. Int J Psychophysiol. 2004;53:217–31.

Wortman CB, Loftus EF, Marshall ME. Sensation and perception. In: Wortman CB, Loftus EF, Marshall ME, editors. Psychology. New York: Mc Graw-Hill; 1992.

Chapter 6
Mind and Space

This chapter describes various approaches to complexity and entropy in the central nervous system that may explain time and space changes in neural synchronization and coherence. These changes in brain complexity are likely the basis for discrete mental states that, through their differences, enable recognition and awareness of the external and internal world. According to this concept, the image of the world emerges as a consequence of creating order arising from nonlinear activities of large groups of neurons. These highly organized nonlinear processes are a consequence of high system complexity that occurs when the system involves a large number of interlinked and simultaneously active neural assemblies and runs in a desynchronized parallel distributed mode that can lead to self-organization. These levels of complexity and entropy within the brain likely present basic code that enables mental and physical space to be connected and corresponding differences and their recognition in mental and physical space to be defined. This approach provides the possibility of studying "neurogeometry" as a geometrical model of the functional architecture of the brain, which, through neural complexities, can reflect the geometry of the external space in the mental space. Within this context, the solution to the binding problem could principally use similar mathematical approaches to those studied in physics, and also include descriptions of how specific observers "define" reality and create observer-specific geometry of the space, such as in the general theory of relativity and other theoretical concepts in physics that take into account the role of the observer in the physical world.

In the well-known book *The Astonishing Hypothesis: The Scientific Search for the Soul*, Crick (Fig. 6.1; 1994) argued that the traditional "Cartesian" concept of the soul as a nonmaterial being must be replaced by a scientific understanding of how the brain produces the mind. Two years before this book was published, Crick and Koch (1992) suggested that a new scientific framework for the study of consciousness comparable to the formulation of quantum mechanics in physics might be needed. They wrote:

> How to explain mental events as being caused by the firing of large sets of neurons? Although there are those who believe such an approach is hopeless, we feel it is not productive to worry too much over aspects of the problem that cannot be solved scientifically or,

P. Bob, *Brain, Mind and Consciousness: Advances in Neuroscience Research*,
DOI 10.1007/978-1-4614-0436-1_6,

Fig. 6.1 Francis Crick

> more precisely, cannot be solved solely by using existing scientific ideas. Radically new concepts may indeed be needed—recall the modifications of scientific thinking forced on us by quantum mechanics. The only sensible approach is to press the experimental attack until we are confronted with dilemmas that call for new ways of thinking.

6.1 Entropy, Complexity, and Brain Space

The word entropy originates from the Greek word *τροπη*, which means "transformation." The term was used for the first time and developed by Carnot in 1824 for the explanation of certain processes during the transformation of different forms of energy. Carnot discovered that in principle any engine cannot produce more energy than is supplied for it to function (the so-called first law of thermodynamics, which is an expression of the conservation of energy). This principle states that energy can be transformed from one form into another, but cannot be created. In addition, he found that the energy input provided for an engine to function must be higher than the output, which means that its effectivity = output/input is lower than 1 (the so-called second law of thermodynamics). This principle implies that any energy transformation in engines, physical systems, chemical processes, or in biological systems leads to energy loss of the output. Carnot called this phenomenon "entropy." He also provided a mathematical formulation of this principle that enables the energy loss during transformation to be calculated (Perrot 1998; Haynie 2001; Feynman et al. 2005).

This evidence is also in agreement with everyday experience of any process, physical, chemical or biological, producing heat energy, because all of these

Fig. 6.2 Erwin Schrödinger

processes are related to macroscopic movements that produce heat due to friction (i.e., friction is the force resisting the relative motion among all moving objects such as solid matter, fluids or gases; e.g., Feynman et al. 2005). However, the question is: what is going on in the microscopic dimension of the small particles of these moving bodies?

A response to this question was proposed by Ludwig Boltzmann in the 1890s, who for the first time formulated the so-called statistical interpretation of entropy. He found that the heat "energy" generated by macroscopic physical, chemical, and biological processes, increases disorder on a microlevel. In his statistical interpretation, Boltzmann proposed that the entropy measures the disorder in an ensemble of microparticles that forms the body (the quantitative expression of the entropy is: entropy = k log D, where k is the so-called Boltzmann constant (k = 1.38065 × 10–23 J/K; and D is a quantitative measure of the disorder in a very large ensemble of the microparticles; Schrödinger 1944). Erwin Schrödinger (1944, Chap. 6) later suggested that entropy might be the statistical tendency of matter to go over into disorder. However, the question remains: what is going on in the microscopic dimension of the small particles that form living bodies?

6.1.1 "What is Life" and Information Entropy

Erwin Schrödinger (Fig. 6.2), who is well known for his seminal works on quantum physics, in 1944, before the discovery of DNA by Watson and Crick, proposed that living organisms create order from disorder. In his famous book *What is Life,*

Schrödinger (1944), in Chap. 6, states: "... living matter, while not eluding the 'laws of physics' as established up to date, is likely to involve 'other laws of physics' hitherto unknown, which however, once they have been revealed, will form just as integral a part of science as the former."

There is evidence that particles constituting living bodies are highly organized into cellular and subcellular structures and that macroscopic entropy produced by a living organism does not emerge directly at the microscopic level of the organism. In this context, Schrödinger (1944) proposed that generating information means to produce order and "negative" entropy, which suggests that increased statistical entropy might mean loss of information. In addition, later scientific evidence indicates that the process of generating information and creating order is not static and in equilibrium (e.g., Prigogine, and Stengers 1984; Haynie 2008). A characteristic feature of this repeated information generation is a spontaneous tendency to create dynamical order that emerges in the form of coherent patterns at various levels from molecules to behavior that is denoted as self-organization (Ashby 1947; Prigogine, and Stengers 1984; Hess and Mikhailov 1994; Haynie 2008).

In 1948, Claude Shannon proposed an interesting and useful concept for the description in mathematical language of some of these processes that occur in the living systems, in the article *A Mathematical Theory of Communication*, which includes his formulation of information entropy (Shannon 1948; Shannon and Weaver 1963). Later research has shown that Shannon's concept of information entropy is compatible with Boltzmann's statistical formulation of the entropy, and stated that the loss of information increases spatial disorder (Scott 2005; Volkenstein 2009). This compatibility is based on the works published by Szilard (1929), Rothstein (1951), and Brillouin (1956), who found an interesting connection between Boltzmann's entropy and Shannon's information entropy (Frieden 2004). The seminal article *On the Decrease of Entropy in a Thermodynamic System by the Intervention of Intelligent Beings*, connecting Boltzmann's statistical entropy with the information theory, was published by Szilard in 1929. In this paper Szilard (1929/1964), using a thermodynamic *Gedankenexperiment*, has shown that an observer who performs the experiment, in order to learn, receives information. Szilard also defined today's "Szilard's limit," which means energy that is transformed into increased thermodynamic entropy, and a physical connection formulated between thermodynamic entropy and information (Szilard 1929/1964; Zurek 1984). A similar *Gedankenexperiment* was later also published by Brillouin in 1956 (Brillouin 1956; Frieden 2004). In addition, Brillouin (1956) found that information connected with a specific physical system that is bound in the system, is related to its entropy. He found that the entropy decreases when information is obtained and the information must be furnished by some external agent whose entropy will increase, i.e., bound information = decrease in entropy = increase in negentropy (i.e., negative entropy), and on the other hand, information loss = increase in entropy = decrease in negentropy, which means that the increase in entropy and loss of information proceed together (Brillouin 1956).

In this context, information loss in communication theory is most frequently defined as an unreceived input to an organism or intelligent machine that is presented in time (Scott 2005; Haynie 2008; Volkenstein 2009). On the other hand, a necessary condition for receiving any information is the ability to recognize. Any message includes information only when it is possible to say "yes" or "no" (1 or 0, which defines the basic information unit—one bit), which means that information is impossible to recognize without a previous intention and context, and this forms a pattern of recognition that is compared with the present time situation. The basic characteristic of the recognition pattern is a clearly defined structure that contains all the elements needed for its repeated recognition (Jain et al. 2000; Duda et al. 2000). An important aspect of the recognition and information receipt is a time needed for the recognition process, because any information processing unit is not able to detect two or more pieces of information at the same time, for example, an artificial neuron in neural networks (Valentine 1989; Dierig 1994; Agre 1997). However, in principle, the same applies to any intelligent machines and organisms, including humans. For example, we cannot respond to two questions at the same time; rather, the responses must be arranged in time sequence from first to second, which reflects the fact that the associative process of thinking and feeling only has a time dimension, even though its contents reflect the quality of space.

In this context, Shannon (1948) reported information entropy to be a measure of temporal disorder quantifying a loss of information (in bits), which is mathematically defined in a similar way to Boltzmann's entropy, but the space is replaced by time. Shannon's concept enables information to be understood as an organized sequence of temporal events that creates meaning with its reception. However, in principle, a temporal structure of individual events may also be random and without a defined meaning, or organized in a temporal pattern that creates context (Balian 2003; Scott 2005; Haynie 2008; Volkenstein 2009).

6.1.2 Information Entropy and Brain Complexity

According to recent findings various mutual relations in the brain are mediated by numerous and rapid changes in brain complexity (Elbert et al. 1994; Friston et al. 1995; Tononi et al. 1996, 1998a; Tononi and Edelman 1998; Sporns et al. 2000; Freeman 1991, 2000, 2001). These changes in complexity are represented by various levels of dependence and/or independence among oscillating neural assemblies with characteristic transient periods of synchronization and integration with a lower level of complexity. There is evidence that these transient synchronized oscillations dynamically linking neurons into assemblies at gamma frequencies (30–100 Hz) are closely associated with sensory processing, attentional selection, and effective sensory–motor integration, and also play an important role in working and long-term memory (Fell et al. 2003; Womelsdorf and Fries 2006, 2007; Jensen et al. 2007; Sauseng and Klimesch 2008; Fries 2009). During recognition and selective attention processing, these dynamical interactions of various independent activities produce neural "recognition patterns" that enable sensory information to be compared with patterns in memory and distinguish differences between these patterns. This process is likely based on repeated comparisons of various competitive neural patterns, representing possible interpretations of the information received during selective attentional processing (Desimone and Duncan 1995; Baars 1988, 2002; Kanwisher 2001). Selected interpretation from this competition and its neural pattern constitute the output of the recognition process, which is subjectively experienced as a mental representation of the interpretation.

With respect to the current evidence regarding attentional processing, it is likely that these complex dynamical interconnections, forming various neural patterns, and comparisons among them, produce "differences" that define the basic "code" for recognition. These differences are likely encoded through specific dependencies or independencies among the neural patterns that define hierarchical scales of identities and differences that are interpreted in the "theater" of the mind as different, mutually dependent or independent objects.

The recognition process determined by hierarchically defined differences is basically dependent on signal spread to sensory cortices and other structures participating in signal processing. In this context, Shannon's principle of communication implies that any event of information propagation and its processing leads to information loss and produces information entropy, which defines the "uncertainty" of the signal for its receiver. This uncertainty in information processing specifically influences the recognition process and increases the probability of the recognition error. The relationship between the information loss (producing "uncertainty" or "ambiguity") and the possibility of recognition suggests that the loss of information during propagation, for example, distinguishing between two people when it is getting dark, may lead to perceptual ambiguity. On the other hand, the same effect of uncertainty may be caused, for example, by two well-perceived, conflicting stimuli, for example, during the Stroop color–word interference task, or during perceptual instability in binocular rivalry, and in other experiments related to the processing of incongruent and conflicting information (Kanwisher 2001). This relationship between information loss and the possibility of recognition implies that the main feature of recognition is the definition of the recognition pattern, which clearly describes the difference between the objects. As mentioned, the "difference" may be significantly decreased or lost during signal propagation, but also during perceptual processing, leading to conflict.

In this context, the missing information before its reception may not only mean a loss of information during signal propagation to sensory cortical areas, but may also be caused by incomplete information integration or binding during brain information processing related to conflicting information. Consequently, the loss of information caused by incomplete information integration and disturbed binding implies spatial separation of several neural activities from dominant, integrated neural processing. It implies that an increase in information entropy might occur because of the disintegration of several information channels and processors, which may determine information loss. On the other hand, as the relationship between Boltzmann's statistical entropy and Shannon's information entropy predicts, the loss of information related to disintegration of processed information increases the spatial disorder of the brain, which may consequently influence deficits in the binding mechanism and other brain functions.

What in fact determines this binding among spatially distant neural assemblies that, as a whole of connected parts, participate in transforming the flow of information, remains a puzzle. However, what is most important for psychology and brain sciences in the relationship between statistical (material) entropy and information entropy is its ability to connect the world of physical bodies—*res extensa*—with the "Cartesian" soul as a nonmaterial being—*res cogitans*—because both these domains are involved and connected in the processes of information exchange.

6.1.3 The Brain–Mind Information Principle and Its Neuroscientific Implications

The relationship between statistical (material) entropy and information entropy enables both the material and the information domains to be described and connects them on the principle of mutual connection between the loss of information, which determines information entropy growth, and increased disorder in the material domain. As a consequence, this principle of mutual connection between the material and information domains permits the concepts of information entropy and statistical entropy to be rigorously applied to brain information processing.

Coordinated brain activity and information processing is based on a series of temporally and spatially organized microstates (i.e., neural binding) and the brain as a physical system has a unique ability to combine neural bits of information into complex patterns of activities that enable mental functions and behavior. A particularly intriguing result of this brain information processing is the process of self-reference based on observing the process and the ability to distinguish between the self and an external world. This ability of the brain's self-organization to coordinate neural microstates and "bind" them into complex patterns of information enables basic terms from thermodynamics and information theory to be used and extrapolated to specific brain information processing. Based on this relationship it is possible to connect a level of binding that defines the time and spatial coordination of neural microstates with information patterns in the brain. This connection specifically means that disturbed binding implies an increase in statistical disorder on a neural level related to various physical and molecular processes that consequently (at a higher level of disarrangement) may appear as functional or structural deficits and may determine neurodegenerative changes in various structures of the brain. However, although this relationship between the brain's physical and information processes is evident, there is an epistemological limitation in understanding these processes. This limitation is necessary, because although there is evidence that mental phenomena are related to various physical, molecular, and neurophysiological processes, it is not possible to suppose that all brain information processes have mental representation strictly defined in time and space. Yet, it is evident and without doubt that large (indefinable) parts of the brain's information content and related physical and information processes are linked to the human consciousness, which may be defined as a specific form of brain information processing. In addition, including the time dimension as a sequence of mental phenomena that produce states of consciousness enables the mind to be defined as a specific set of mental states represented by various states of consciousness and their connections, which provide associative chains. In this context, it is possible to connect physical and molecular processing linked to neural microstates with the level of information processes in the human mind. Based on this relationship, it is possible to define brain entropy as a level of disorder that reflects the deficits in the binding of neural microstates, which consequently defines "entropy" of the mind as a specific level of disorder that is subjectively experienced and reflected in human behavior.

This physical–information relationship enables a basic principle of brain–mind information exchange to be defined that is based on the relationship between brain physical processes and information processing in the human mind. The relationship between statistical (material) entropy in the brain and information entropy related to information content in the human mind, therefore, represents and constitutes the basic brain–mind information principle, which specifically states that the loss of information during brain information processing implies increased disorder (and entropy) of the brain and mind.

Although the brain–mind information principle is a consequence of general physical laws, it represents at the same time an experimentally testable hypothesis for brain–mind sciences. Using this principle it is possible to connect psychophysiological and psychometric methods and experimentally assess the complexity and other measures reflecting order, disorder, and entropy in the physical domain as well as in the psychological domain, and find relationships between them, as several recent data indicate (Bob 2007; Bob et al. 2009a,b, 2010a,b; Bob and Svetlak 2011).

The relationship between statistical (material) entropy and information entropy in principle enables the mathematical methods of the complexity theory (and other parameters reflecting order and disorder) to be applied to psychophysiological measures (such as EEG, ECG or electrodermal activity [EDA], fMRI, and other measurement tools), and use them together with methods that monitor the psychological state, such as psychometric measures of mental disorders, for example, dissociative symptoms, that also enable a level of order and disorder to be quantified. From this point of view the brain–mind information principle constitutes a relationship between psychophysiological and psychometric measures on a rigorous physical basis that does not only represent an analogy between the complexity of the mind and brain, but is also a logical consequence of basic principles connecting the two domains as parts of the same world, which in principle indicates that an information change in one domain implies an information flow to the other.

6.1.4 Brain–Mind Information Principle and Some Examples of Its Application

A typical example of the application of the brain–mind information principle is possible to find in research on schizophrenic patients. The brain–mind information principle states that the loss of information during brain information processing implies increased disorder (and entropy) of the brain and mind. In the context of schizophrenia, the brain–mind information principle predicts that the loss of information typical for misinterpretation related to delusive thoughts implies disorder and increased statistical randomness in the microscopic spatial domain of the brain. On the other hand the loss of information also predicts increased randomness in the temporal domain of the mind, for example, pseudo-randomness in the association flow of schizophrenic patients and other temporal discontinuities of mental experience, such as amnesia, depersonalization or derealization, and other symptoms.

This consequence of the brain–mind information principle is in agreement with findings documenting the disorder in the brain's spatial domain in schizophrenic patients. There is evidence that spatiotemporal binding and synchronization, mainly in the gamma range related to brain information processing, is significantly affected in schizophrenia (Tononi and Edelman 2000; Peled 1999; Lee et al. 2003; Ford and Mathalon 2008; Uhlhaas et al. 2008; Uhlhaas and Singer 2010).

On the other hand disturbed order related to information loss in the temporal domain is documented in studies of the associative process, and other studies focused on the discontinuities in schizophrenic thinking and its pseudorandom behavior as observed in word associations (Jung 1909; Kent and Rosonoff 1910; Moran et al. 1964; Shakow 1980; Goldberg and Weinberger 2000), impaired verbal fluency (Allen et al. 1993; Himelhoch et al. 1996; Vinogradov et al. 2002), and textual analyses of the semantic processing (Manschreck et al. 1979, 1981; Hoffman et al. 1982; Goldberg and Weinberger 2000), indicating deficits in the organization of semantic memory in schizophrenia (Davis et al. 1995; Paulsen et al. 1996; Vinogradov et al. 2002).

Another experimental prediction of the brain–mind information principle is focused on the experimental research of stress and cognitive conflicts that lead to the loss of information because of information interference. In this context, application of the brain–mind information principle predicts microscopic spatial disorder in the brain and temporal disorder of information in the mind. There is evidence that a high level of stress leads to extreme cognitive and emotional conflict, which produces dissociation as a disorder in the temporal memory domain (Kenardy et al. 2007; Payne et al. 2006; Brewin 2007). There is preliminary evidence that dissociative symptoms are related to global disruptions in spatial binding and synchronization of the brain in schizophrenic patients (Bob et al. 2010a,b), and increased complexity in healthy people (Bob and Svetlak 2011).

In addition, the brain–mind information principle demonstrates dissociation to be a consequence of the information loss related to an inability to integrate some mental content into consciousness, which determines disorder in the temporal information domain of the mind. As well as those typical events of information loss, it is possible to find dissociation in various experimental and clinical situations, such as in hypnosis, during stressful tasks leading to conflict or information overload, and in various clinical cases. In this context some presentations of dissociation in its severe forms, such as dissociative identity disorder or in some cases of hypnosis, could represent a logical consequence of the basic principles of nature that predict disorder and pseudo-randomness in information integration within the temporal domain of the mind, as a consequence of specific psychological and neurobiological changes leading to information loss in brain information processing.

Other applications of the brain–mind information principle to psychological and brain sciences could help to explain specific spatial and temporal changes in various psychopathological states that may occur because of a loss of information in the context of human communication. It is mainly because missing components of some patterns of human communication or specific forms of abuse, such as neglect, may lead to a loss of specific information needed for normal development of the brain and psychological integrity. This information loss on a general level involves specific deficits and disorder on a spatial scale of the brain and the temporal information domain of the mind.

From this point of view the brain–mind information principle within the framework of the general principles of self-organization could provide an explanation for the basic concept of social neurosciences, which is in agreement with Schrödinger's (1944) statement that life needs information in its various forms to balance the statistical tendency of matter to go from order to disorder. This principle is in agreement with findings that not only harmful information or damage to the brain leads to pathological consequences, but also information missing on social contacts may lead to brain disorder (for example, Teicher et al. 2003, 2006). Therefore, missing components of some human communication patterns, or specific forms of abuse, such as neglect, leading to a loss of specific information needed for normal development of the brain and psychological integrity, thereby leading to specific deficits and disorder on a spatial scale in the brain and the temporal information domain of the mind. From this point of view, the principle of information entropy explains the evidence of social neurosciences within the framework of the general principle of self-organization, which states that life needs information in its various forms to balance the statistical tendency of nature to go from order to disorder.

Also worthy of attention is the prediction of the brain–mind information principle for the interruption of consciousness in anesthesia related to an almost complete loss of information, which also predicts disorder in the spatial domain of the brain and the temporal domain of the mind. This prediction is in agreement with recent evidence in anesthesiology indicating that the anesthesia-mediated interruption of consciousness may be related to an "unbinding" mechanism across various parts of the brain (Mashour 2004, 2008; Alkire 2008).

For experimental testing of the hypotheses, it is possible to use a wide spectrum of methods of mathematical physics and information theory, such as time series analysis applied to psychophysiological measures, which enables complexity to be analyzed or measure of the disorder in the system to be specifically provided, such as Fisher information and other methods of complexity and nonlinear dynamics (Kantz and Schreiber 1997; Frieden 2004). On the other hand, there is an enormous amount of very useful and well-elaborated methods of psychological testing that are able to provide statistically reliable quantitative data (for example, see Kline 2000; Groth-Marnat 2003).

The critical epistemological aspect of the brain–mind information principle is that although it is general, it may be specifically useful in various fields of research.

On the other hand, the brain–mind information principle is falsifiable because it is not able to explain the binding problem, which is not within its terminological framework. In this context, application of the brain–mind information principle in this field of research, although it is limited, it may help to integrate various findings in the field in a meaningful way, it is compatible with substantial findings in the field, and using this principle may help to establish a novel brain–mind information theory that might be useful in making specific, experimentally testable predictions and defining methods for their performance.

As mentioned above, application of the brain–mind information principle in psychology, psychophysiology, and cognitive neurosciences is capable of providing specific experimental predictions representing regular implications of the theory. In addition, application of the concept could in principle enable psychophysiology, cognitive neurosciences, and experimental psychology to be integrated into the domain of interdisciplinary physical sciences based on rigorous applications of the mathematical theory and experimental data acquisition. In this context, the integration may enable reasonable and meaningful utilization of some mathematical methods and rigorous testing of the theoretical predictions formulated in mathematical language using statistical and numerical calculations performed on experimental data. Although in principle the psychological data are not as accurate as physical measurements, these data do not change anything with regard to the promising perspective of utilizing the theoretical and experimental principles of physical sciences in the field of neuroscientific and psychological research, and connecting both parts of the "Cartesian" universe using scientific methods.

6.2 The Binding Problem and Some Principles of the Theory of Relativity

"Neurogeometry," proposed as a geometrical model of the functional architecture of the primary visual cortex and its pinwheel structure (Petitot 2003), provides a motivating and nontraditional contribution to the solution of the binding problem. The problem is to understand how the internal geometry of the visual cortex can produce the geometry of the external space. Petitot, using this analytical approach, found that the specific mathematical condition of integrability reflecting the "connectivity" of the geometrical structure represents a neurogeometrical condition of the binding mechanism among distributed neurons that allows understanding of the typical Gestalt phenomena. Petitot also suggested that a synchronized wave of activity propagating in the functional architecture might be geometrically equivalent to the integration of a specific differential equation describing the internal brain geometry that perspectively enables the problem of binding to be understood as a physical–geometrical process and closely connects the neurosciences and physical sciences (Petitot 2003).

Within this context a solution to the binding problem could principally be explained by a similar mathematical approach to that used in the general theory of relativity,

Fig. 6.3 Bernhard Riemann

which, similar to the concept of neurogeometry, uses Riemannian geometry of the curved space (Bob 2009). This geometry was first described by Bernhard Riemann (Fig. 6.3) in the 1850s, who had the original idea to use a specific collection of numbers describing the qualities of every point in space, which enables us to characterize how much of the space is flat or curved (the so-called Riemannian metric). Later, in 1915, Albert Einstein used the Riemannian metric for his formulation of the general theory of relativity (Laue 1950). This geometrical approach enables different observers' viewpoints to be described, leading to different measures such as typical changes of the space metric, for example, the length between two points may be measured differently by two different observers (Einstein 1916/2005; Penrose 2004). In the context of neurogeometry it means that, for example, two areas of the visual cortex that seem to be distant from the external point of view might have zero distance from the viewpoint of the internal geometry.

The Riemannian geometry (or metric) may be open, flat, or closed. This specific curvature means that a defined length in flat space (for example, 1 m), will be observed and measured in the space (from another point of view) with the open geometry being longer (than 1 m; or in an extreme case unlimited), whether "the same length" observed and measured in the space with the closed geometry will be shorter (than 1 m; or in an extreme case, zero).

In addition, there is evidence that these geometrical principles have physical reality and are not only the product of abstraction. Very interesting examples are the physical changes in moving bodies that occur when their velocity is close to the speed of light. When the observed body significantly changes its velocity approximating the speed of light, the observer registers the distance contraction, time slowing, and increased mass and energy. In addition, the effect of observer movement with significant acceleration leading to accumulation of high mass and energy is likely equivalent to the influence of a gravitational field, which presents Einstein's equivalency principle as a basic postulate of the general theory of relativity (Einstein 1916/2005; Pauli 1958; Beiser 1995). For example, high mass and energy accumulated in a black hole cause similar effects of distance contraction and time dilatation (Einstein 1916/2005; Penrose 2004).

Application of this geometrical concept to the brain may imply that the brain metrics from the point of view of the observer (with the brain as an object), may not be the same as the brain's interpreter, which enables consciousness. From this point of view, the binding problem may be resolved by a nonexistent distance among the parts of the brain from the point of view of the brain's interpreter, because of the different space metrics that enable brain synchronization and coherence (Bob 2009).

In this context, closed geometry could play a role in brain functions during very short periods of its functioning (to the order of milliseconds, or less) in which attention is fully focused on internal processes and is closed to external stimuli. Those very short periods of time during discontinuous flow of "flashing" attention may cause the brain's internal geometry to be closed with respect to outside stimuli, which could enable neural binding with "zero distance" among simultaneously active groups of neurons from the point of view of the brain's interpreter. This "zero" distance as a seat of the mind ("Cartesian theater") must not be strictly zero in length, but in principle may have a minimum length that occurs in the universe. As a potential "seat of the mind," this minimum length, according to recent quantum physics, could be the so-called Planck's length ($\approx 10^{-35}$ m), which is based on the constants c, G, and $\hbar$, where c is the speed of light in a vacuum, G is the gravitational constant, and $\hbar$ is the reduced Planck constant (Planck's length $= \sqrt{(\hbar G/c^3)}$). Although at this time the ideas about brain-specific geometry are predominantly speculative, the different perspectives of external and internal observation may play a role, and imply that the spatial distance between two points in the brain must not be the same from the viewpoint of an external observer in comparison to an internal observer, which is the brain's interpreter or the subject's mind itself.

6.3 Conscious Observers and Quantum Physics

The analogical problem of observation, which plays a part in "reality formation," emerges explicitly in quantum theory. Heisenberg (1958, p. 22) was of the opinion that the transition from the "possible" to the "actual" in quantum theory takes place

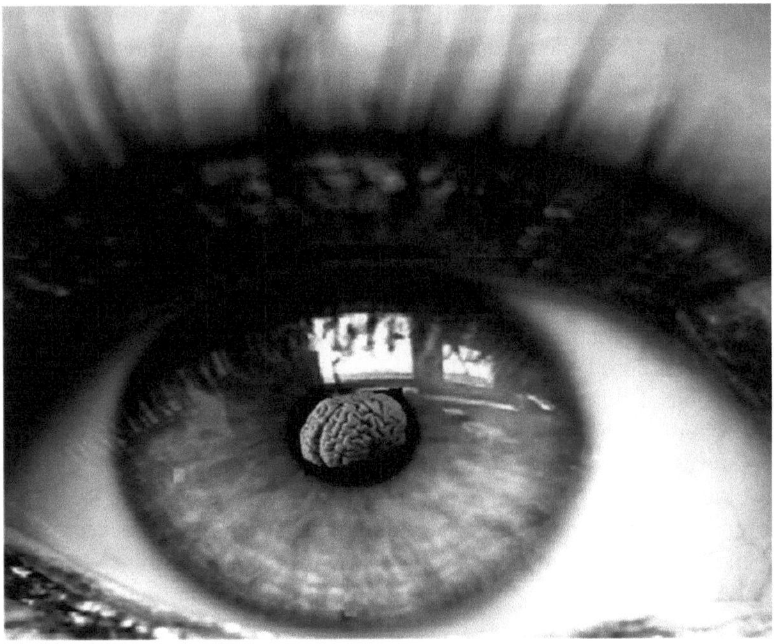

Fig. 6.4 Three worlds and the hierarchy of observers

during the act of observation. It was expressed more explicitly by Bohr (1958, p. 81), who briefly summarized a key consequence of the quantum theory in a few words: "In the great drama of existence we ourselves are both actors and spectators."

A similar view was put forward by von Neumann (1955, p. 418), who emphasized that the interpretation of quantum theory requires the existence of a subjective (i.e., conscious) observer, whose mental activity influences physical processes. In this context, von Neumann divided the world into three parts (I, II, and III). According to these divisions, part I is everything up to the retina of the observer, part II includes the observer's retina, nerve tracts, and the brain, and part III is the observer's abstract "ego" (von Neumann 1955, p. 420) (Fig. 6.4). In addition, von Neumann considered a case in which the observer forgets the information, and found that this process also leads to an increase in entropy (Neumann 1955, p. 417–437; Brillouin 1956, p. 157).

Bass (1975) developed a detailed description of the quantum process during the act of observation. He proposed a semirealistic neurochemical model that describes the entry of a datum into the consciousness of an observer, which influences the excitation of a nerve cell in the observer's central nervous system. Bass suggested that the mind can induce muscular movements by choosing to note or not to note a relevant datum originating from specialized elements of the nervous system.

Mould (1995, 1998) also suggested some interesting arguments in favor of the quantum description of the mind, and proposed that conscious brains, similar to atoms or black holes, are parts of the quantum mechanical universe. He proposed specific conditions and rules that define the conscious brain as an inside observer, in contrast to a conscious observer existing outside the system, and suggested that there is a neurological mechanism that responds to the presence of an inside observer, with experimentally testable consequences.

In principle, Stapp (2001) proposed a similar way of thinking, reflecting that physical processes in human brains are a necessary condition for the study of the mind–brain interaction. Stapp, in agreement with an interpretation proposed by von Neumann, assumes that an interaction between the objective quantum universe and human consciousness leads to a sudden change that brings the objective physical state in line with the subjective experience of information. In the context of quantum brain information processing, a specific role played by some quantum phenomena in attentional processing has been proposed that suggests that an attentional mechanism related to the observer's mental activity may present a quantum phenomenon analogous to the so-called quantum Zeno effect. The Zeno effect (a parallel to Zeno's paradox regarding the impossibility of motion) can be defined as a phenomenon in which the time evolution of a quantum object can be suppressed and the system repeatedly returns to its initial state because of repeated measurement, which does not allow the system to finish its evolution, forcing it to start its evolution again, for example, in the decay of unstable particles and other experimental arrangements (Sudarshan and Misra 1977; Igamberdiev 2007; Maniscalco et al. 2008). Stapp proposed that the Zeno effect during quantum brain processing could explain several phenomena related to keeping attention focused on a task. In line with psychological data (Pashler 1998; Stapp 2001) a time between 100 and 300 ms seems to be needed to fix attention without interruption and, for example, increased effort that interrupts attentional focus may start the process again (Stapp 2001, 2004, 2005; Schwartz et al. 2005). Further applications of quantum physics in psychology and neuroscience were also proposed by Schwartz et al. (2005), who suggested that quantum theory, which is not limited only to the research of local interactions, could provide a useful theoretical framework for experimental neuroscience that may perspectively also enable us to understand more complex neurobiological mechanisms that emerge in cognitive functions related to large-scale integration in the brain.

6.4 The Binding Problem and Quantum Nonlocality

Quantum nonlocality (or entanglement) is a physical process that enables instantaneous correlations between interacting, but spatially separated quantum subsystems. For example, a pair of two spatially distant particles in a nonlocal (or entangled) state can be used as an information channel that can enable information to be transmitted instantaneously and computational and cryptographic tasks to be performed, which is not possible in the classical nonquantum systems (d'Espagnat 1976; Bell 1987; Penrose 1994; Nielsen and Chuang 2000). Although this information is intuitively controversial, the study of information processing in quantum systems is a subject in quantum computation theory that has the specific aim of developing quantum computers (Haroche and Raimond 1996; Nielsen and Chuang 2000).

A historical origin of the term "quantum nonlocality" is the so-called Einstein–Podolsky–Rosen paradox (EPR paradox) (Einstein et al. 1935; d'Espagnat 1976; Penrose 1994; Wheeler and Zurek 1983). The EPR paradox, an important topic of quantum physics, consists of a mathematically described thought experiment of two interacting microscopic systems (such as two interacting electrons or photons). For example, two electrons may be in one "box" (such as an atomic orbital) only under the condition of a different spin number (↑↓; i.e., ½ or −½—the so-called Pauli principle). This spin number does not exist as an independent variable, but it is defined by a specific act of measurement, for example, in a magnetic field. When measurement using the magnetic field is performed on the one electron of the pair, the electron has a degree of freedom (it may choose) with respect to the direction of the field, i.e., it may choose between spin ½=↑ or −½=↓. When the electron selects ↑, then the second electron of the pair is instantly influenced by this choice and has a reciprocal spin value, ↓. The only limitation of this experiment is that both electrons must be interacting in one quantum state (i.e., in one "box"). However, there is no condition stipulating how large the box can be, and there is evidence that the interacting pair (of photons) may be spatially separated and that the metaphorical "box" (representing interaction between the particles) may be, for example, a few kilometers (or miles) long (Aspect et al. 1982; Wheeler and Zurek 1983; Penrose 1994; Aspect 1999). According to quantum theory, the second electron responds to the measurement performed on the first electron immediately (and instantly) by reciprocal orientation of the spin independent of the distance between the electrons (which defines the length of the "box") (d'Espagnat 1976; Albert 1992; Ballentine and Jarrett 1997).

It results in a paradox, whereby the measurement of a physical quantity in one system affects the measurement of a physical quantity in another, spatially separated system immediately. In this context, the paradox contradicts the theory of relativity, which defines that the limiting velocity of any signal is the velocity of light. Because of this paradox, Einstein denied accepting quantum mechanics as a "real" and complete theory, but later research has shown that this paradox (also called the quantum nonlocality) really exists and that quantum mechanics violates classical intuitions (Einstein et al. 1935; d'Espagnat 1976; Wheeler and Zurek 1983; Bell 1987; Penrose 1994).

The nonlocality, in principle, means entanglement between two initially interacting micro-objects across a distance instantly (with zero lag correlation). Although the phenomenon is very curious, technologies using quantum entanglement are currently being developed. For example, in so-called quantum computation the entangled quantum states may allow parallel processing and certain computations may be performed much more quickly than with classical computers (Gottesman and Chuang 1999; Haroche and Raimond 1996; Nielsen and Chuang 2000). Parallel processing in the quantum computers may also run on the principle of the nonexistent distance between spatially distributed processors, analogous to the brain functions, which are thought to have specific integrative abilities to bind distributed information independently of distance among oscillating neural populations.

Based on this principle, Marshall (1989) proposed a nonconventional concept for the binding of distributed information in the brain; in fact, there exists neither a classical physical structure nor a neurophysiological substrate suitable for explaining consciousness. Instead, Marshall focuses his attention on the "quantum wholeness" of initially interacting, but (in the future) spatially distributed, subsystems. In quantum reality, the nonlocal instant correlations between spatially distributed parts of a system are possible, independent of any connecting signal between them that could explain the unity of consciousness during distributed information processing.

Beck and Eccles (1992) also proposed a well-known contribution to the role of quantum mechanics in brain processes. They studied how collective and synchronized activities of many neurons are related to neurotransmitter release and how through this collective mode of behavior mental intention may become neurally effective by momentarily increasing the probability of exocytosis. In their research they found that neurotransmitter release from the presynaptic part of the neuron into the synaptic cleft has a probability of less than 1. They interpreted this probability as a consequence of the quantum "tunneling effect," which enables a particle to overcome an energetic barrier that is higher than the energy of the particle (i.e., a phenomenon that is impossible to observe in our macroscopic world described by classical mechanics).

Analogous to the theory of relativity, the different perspectives between external and internal observation and experience may imply that the spatial distance between two points in the brain must not be the same from the point of view of the external observer in comparison to the internal observer, which is the subject's mind itself, and these effects could be understood to have some connection with changes in space metrics related to quantum entanglement and other physical processes in the brain.

6.5 Brain and Quantum Gravity

In the physics of the twentieth and early twenty-first century it has been found that many physical phenomena can be described well by quantum mechanics and the theory of relativity. Both theories are able to describe physical reality, but even though there have been many attempts to find a connection between the two theories since the advent of modern physics, a unification has not been reached (Kragh 1999; Callender and Huggett 2001; Penrose 1994, 2001; Smolin 2004). In this context, recent quantum gravity physics primarily refers to an area of research in which several approaches to the unification of quantum physics with general relativity have been proposed, but none of them has been entirely successful. One of these approaches was suggested by Penrose (1989, 1994, 2001), a system unifying quantum mechanics with Einstein's general theory of relativity that also takes into account brain functions and consciousness. This approach to the physics of consciousness, within the framework of general findings that might be important for the theory of the universe, has been developed by a group of researchers who studied some of the experimental predictions of the theory (Hagan et al. 2002; Hameroff and Penrose 1995, 1996; Penrose 1989, 1994, 2001; Penrose and Hameroff 1995; Woolf and Hameroff 2001).

The crucial problem emphasized in Penrose's approach focuses on the issue of measurement during microphysical processes in quantum mechanics (Penrose 1994). In a classical solution by Bohr (called the Copenhagen interpretation) a collection of possibilities (e.g., all possible trajectories connecting two spatially distributed points) in the development of a system (characterized by the wave function Ψ) is reduced to one macroscopic actualized possibility (e.g., one of a set of possible trajectories) by measurement or observation performed on the quantum system (Laurikainen 1988; Penrose 1994; Wheeler and Zurek 1983). This process emerges in the transition from the quantum reality of possibilities and uncertainty (the world of microphysics), characterized by the wave function Ψ, to the macroscopic world of things, characterized by "sharpness," with defined localization in space and time. Penrose (1994) called this process subjective reduction of quantum possibilities. This classical concept of reduction of the wave function led to the well-known Schrödinger's cat paradox. This paradox demonstrates the conflict between quantum theory and macroscopic observations performed on the quantum mechanical system. In a thought experiment, a living cat is placed into a box along with a bottle containing a poison. In the bottle is a very small amount of a radioactive material. If even a single atom of the radioactive material decays during the experiment, a relay mechanism with a radioactive detector will trip a hammer, which will, in turn, break the bottle containing the poison and kill the cat. The observer cannot know whether or not an atom of the substance has decayed and whether the cat is killed. According to the Copenhagen interpretation, the cat is in a superposition of states and is both dead and alive (Penrose 1994, 2004; Wheeler and Zurek 1983). When the observer opens the box, the superposition is lost and the cat becomes dead or alive. According to Penrose, a solution to this problem can be found in the theory of quantum gravity, which might explain the spontaneous (or objective) reduction of the superposition determined by the inner process of a quantum system that is not caused by measurement or observation (Penrose 1994, 2001, 2002). This reasoning is to resolve the paradox that the creation of the world is determined by the process of observation. Instead of this concept, Penrose suggested a mechanism of objective reduction, which means that the process of actualization from the collection of quantum possibilities is not only determined by observation and the measurement process, but may also represent an objective physical process that can spontaneously exist independently of the observer. To possibly resolve the paradox of observation Penrose developed a proposal related to the problem of gravitation in quantum physics.

The reasoning is that the Schrödinger's cat paradox may be caused by an often underestimated issue that gravitation in the microworld is very low and in the classical theory of quantum mechanics it is not necessary to take it into account. An interesting proposal regarding how it is possible to include Newton's concept of gravitation into the framework of quantum physics was put forward for the first time by Karolyhazy (1966). Karolyhazy treated the problem of the reduction of the wave function in quantum theory from a new standpoint in which he combined Heisenberg's uncertainty relations with gravitation and using these relations calculated quantitative limitations of the "sharpness" of the space–time structure. The second step of his approach was incorporation of the resulting uncertainty of the space–time structure into usual equations characterized by the wave function Ψ, which is used in quantum physics for description of particle wave objects. Karolyhazy found that including Newton's law of gravitation into the quantum mechanical description might play a specific role when we take into account large ensembles of microparticles that may also be described using the quantum mechanical wave function. In his description, Karolyhazy (1966) found that a large, isolated ensemble of microparticles described as a whole by the wave function survives only for a very short period of time. The system then spontaneously collapses into the usual order of macroscopic objects, without any of the measurement or observation that is postulated by the Copenhagen interpretation, or in principle a similar theory of decoherence, which postulates that an outside stimulus might be a cause of quantum collapse from possibilities (i.e., the transition from the potentiality of all of the possible places the particles may be) into actuality where they occur (Wheeler and Zurek 1983; Penrose 1994; Schlosshauer 2005). In this context the result of a measurement is not only dependent on the way the measurement is performed, but also on some intrinsic (objective) criteria of the measured system and its spontaneous tendency to develop over time. This view in principle represents a compromise between creating reality by observation, which according to the theory of relativity and quantum physics emphasizes the specific and important role of the observer in the physical processes, and mathematical and physical laws, which represent an inherent part of the universe.

In this historical and philosophical context the mathematical formulation of Penrose's ideas is linked to the theory of "Newtonian quantum gravity," which represents the quantization of gravity in the classical (nonrelativistic) space and time (Ghirardi et al. 1990). Objective reduction, according to Penrose, is linked to noncomputability, which means that gravitational phenomena in principle may have the characteristics of spontaneity and unpredictability similar to what we can see in self-organized systems. What is typical of the systems with self-organized behavior is that they are not strictly predictable, because changes in their behavior are not only determined by outside stimuli, because specific internal dynamics that cannot be included in the objective description may also influence their behavior. From the philosophical point of view, the spontaneous collapse of potentialities (or the objective reduction) that occurs independently of the act of "subjective" observation or measurement, can be interpreted as a nonpredictable spontaneous expression of the underlying mathematical world of potentialities—including physical laws—in the physical world of things.

Penrose (1994) links the objective reduction with the mathematical theorem of noncomputability and unpredictability formulated by Kurt Gödel in 1931. Gödel's incompleteness theorem states that in any formal axiomatic system using the regular rules and axioms, some propositions developed within this system cannot be proven true or false, which means that within the framework of this axiomatic system they are true and false at the same time (Boyer 1968; Chaitin 1982; Rucker 1982; Penrose 1994; Hofstadter 1999; Nagel, and Newman 2002). Because physics uses mathematics as its main tool for the quantification of physical relations, Gödel's incompleteness theorem implies that it is not possible to find a rigorous system of mathematical axioms that would be without any paradox, also for physical theory. The theorem likely represents a substantial limitation of physical sciences, which from this point of view cannot find a theory of everything, based on mathematical axioms, that could explain the universe as a whole.

In agreement with the role of objective reduction in the physical world, Penrose (1994) proposed that on the same principle that enables creation of the external physical world, an inner psychological world may also exist. In this context he also suggested that objective reduction could specifically operate in brain functions and linked this objective reduction process to quantum gravity. Because quantum gravity, as an extension of quantum physics, includes nonlocal instantaneous connections of the interacting quantum objects—the so-called quantum entanglement—this theoretical concept may connect unpredictability related to the system of self-organization with entanglement of the distributed parts of the system, which may help to resolve the binding problem of synchronized distributed neurons that integrate information during brain information processing (Crick 1994; Crick and Clark 1994; Penrose 1994, 2001; Woolf and Hameroff 2001). Based on these assumptions Penrose developed the model of biological coherent quantum states that was initially proposed by Fröhlich (1968, 1970, 1975; see also Marshall, 1989). These coherent quantum states, according to Fröhlich, are caused by electron conformational dynamics of proteins of the neural cytoskeleton (Penrose 1994) and are linked to inter- and intra-cellular communication in the central nervous system (Zaccai et al. 1998). Information distributed in the brain, according to this hypothesis, may be "nonlocally" linked through electron conformational dynamics of microtubule substructures (Penrose 1994, 1997).

Although this hypothesis is very interesting according to its critics (for example, see Tegmark 2000), it is not clear which neurobiological mechanisms enable the survival of the hypothetical coherent quantum states and protect them against decoherence for the time needed for their specific biological functioning, which in principle could enable explanation of neural binding and the synchronization of the distributed neural assemblies.

References

Agre PE. Computation and human experience. Cambridge: Cambridge University Press; 1997.

Albert D. Quantum mechanics and experience. Cambridge, MA: Harvard University Press; 1992.

Alkire MT. Loss of effective connectivity during general anesthesia. Int Anesthesiol Clin. 2008;46:55–73.

Allen HA, Liddle PF, Frith CD. Negative features, retrieval processes, and verbal fluency in schizophrenia. Br J Psychiatry. 1993;163:769–75.

Ashby WR. Principles of the self-organizing dynamic system. J Gen Psychol. 1947;37:125–8.

Aspect A. Bell's inequality test: more ideal than ever. Nature. 1999;398:189–90.

Aspect A, Dalibard J, Roger G. Experimental test of Bell's inequalities using time- varying analyzers. Phys Rev Lett. 1982;49:1804–7.
Baars BJ. A cognitive theory of consciousness. Cambridge: Cambridge University Press; 1988.
Baars BJ. The conscious access hypothesis: origins and recent evidence. Trends Cogn Sci. 2002;6:47–52.
Balian R. Entropy – protean concept. Poincaré Seminar. 2003;2:119–45.
Ballentine LE, Jarrett J. Bell's theorem: does quantum mechanics contradict relativity? Am J Phys. 1997;55:785–92.
Bass L. A quantum-mechanical mind-body interaction. Found Phys. 1975;5:159–72.
Beck F, Eccles JC. Quantum aspects of brain activity and the role of consciousness. Proc Natl Acad Sci USA. 1992;89:11357–61.
Beiser A. Concepts of modern physics. 5th ed. New York: Mc Graw Hill; 1995.
Bell J. Speakable and unspeakable in quantum mechanics. Cambridge: Cambridge University Press; 1987.
Bob P. Hypnotic abreaction releases chaotic patterns of electrodermal activity during dissociation. Int J Clin Exp Hypn. 2007;55:435–56.
Bob P. Quantum science and the nature of mind. J Mind Behav. 2009;30:1–14.
Bob P, Svetlak M. Dissociative states and neural complexity. Brain Cogn. 2011;75:188–95.
Bob P, Susta M, Gregusova A, Jasova D. Dissociation, cognitive conflict and nonlinear patterns of heart rate dynamics in patients with unipolar depression. Prog Neuropsychopharmacol Biol Psychiatry. 2009a;33:141–5.
Bob P, Susta M, Chladek J, Glaslova K, Palus M. Chaos in schizophrenia associations, reality or metaphor? Int J Psychophysiol. 2009b;73:179–85.
Bob P, Palus M, Susta M, Glaslova K. Sensitization, epileptic-like symptoms and local synchronization in patients with paranoid schizophrenia. Prog Neuropsychopharmacol Biol Psychiatry. 2010;34:143–6.
Bohr N. Atomic physics and human knowledge. New York: Wiley; 1958.
Boyer CB. History of mathematics. New York: Wiley; 1968.
Brewin CR. Autobiographical memory for trauma: update on four controversies. Memory. 2007;15:227–48.
Brillouin L. Science and information theory. New York: Academic Press; 1956.
Callender C, Huggett N. Physics meets philosophy at the Planck scale. Cambridge: Cambridge University Press; 2001.
Chaitin GJ. Godel's theorem and information. Int J Theor Phys. 1982;22:941–54.
Crick F. The astonishing hypothesis. The scientific search for the soul. London: Simon and Schuster; 1994.
Crick F, Clark J. "The astonishing hypothesis". An interview. J Conscious Stud. 1994;1(1): 17–24.
Crick F, Koch C. The problem of consciousness. Sci Am. 1992;267(3):153–9.
Davis AV, Paulsen JS, Heaton RK, Jeste DV. Assessment of the semantic network in chronic schizophrenia. J Int Neuropsychol Soc. 1995;1:132.
Desimone R, Duncan J. Neural mechanisms of selective visual attention. Annu Rev Neurosci. 1995;18:193–222.
d'Espagnat B. Conceptual foundations of quantum mechanics. 2nd ed. Reading MA: Benjamin; 1976.
Dierig S. Extending the neuron doctrine: Carl Ludwig Schleich (1859–1922) and his reflections on neuroglia at the inception of the neural-network concept in 1894. Trends Neurosci. 1994;17:449–52.
Duda RO, Hart PE, Stork DG. Pattern Classification. New York: Wiley-Interscience; 2000.
Einstein A. Relativity the special and general theory. Pi Press: New York (Originally published 1916); (1916/2005).
Einstein A, Podolsky B, Rosen N. Can quantum-mechanical description of physical reality be considered complete? Phys Rev. 1935;47:777–80.
Elbert T, Ray WJ, Kowalik ZJ, Skinner JE, Graf KE, Birbaumer N. Chaos and physiology: deterministic chaos in excitable cell assemblies. Physiol Rev. 1994;74:1–47.

Fell J, Fernández G, Klaver P, Elger CE, Fries P. Is synchronized neuronal gamma activity relevant for selective attention? Brain Res Rev. 2003;42:265–72.
Feynman RP, Leighton RB, Sands M. The Feynman lectures on physics. San Francisco: Pearson/Addison Wesley; 2005.
Ford JM, Mathalon DH. Neural synchrony in schizophrenia. Schizophr Bull. 2008;34:904–6.
Freeman WJ. The physiology of perception. Sci Am. 1991;264:78–85.
Freeman WJ. Mesoscopics neurodynamics: from neuron to brain. J Physiol Paris. 2000;94: 303–22.
Freeman WJ. Biocomplexity: adaptive behavior in complex stochastic dynamical systems. Biosystems. 2001;59:109–23.
Frieden BR. Science from fisher information. Cambridge: Cambridge University. Press; 2004.
Fries P. Neuronal gamma-band synchronization as a fundamental process in cortical computation. Annu Rev Neurosci. 2009;32:209–24.
Friston KJ, Tononi G, Sporns O, Edelman GM. Characterising the complexity of neuronal interactions. Hum Brain Mapp. 1995;3:302–14.
Fröhlich H. Long range coherence and energy storage in biological systems. Int J Quant Chem. 1968;2:641–64.
Fröhlich H. Long range coherence and the actions of enzymes. Nature. 1970;228:1093.
Fröhlich H. The extraordinary dielectric properties of biological materials and the actions of enzymes. Proc Natl Acad Sci USA. 1975;72:4211–5.
Ghirardi GC, Grassi R, Pearle P. Relativistic dynamical reduction models: general framework and examples. Found Phys. 1990;20:271–316.
Goldberg TE, Weinberger DR. Thought disorder in schizophrenia: a reappraisal of older formulations and an overview of some recent studies. Cogn Neuropsychiatry. 2000;5:1–19.
Gottesman D, Chuang I. Demonstrating the viability of universal quantum computation using teleportation and single-qubit operations. Nature. 1999;402:390–3.
Groth-Marnat G. Handbook of psychological assessment. Hoboken, N.J.: John Wiley and Sons; 2003.
Hagan S, Hameroff SR, Tuszynski JA. Quantum computation in brain microtubules: decoherence and biological feasibility. Phys Rev E. 2002;65:061901–9101.
Hameroff SR, Penrose R. Orchestrated reduction of quantum coherence in brain microtubules: a model for consciousness. Neural Netw World. 1995;5:793.
Hameroff SR, Penrose R. Orchestrated reduction of quantum coherence in brain microtubules: a model of consciousness. In: Hameroff S, Kaszniak A, editors. Toward a science of consciousness. Cambridge, MA: MIT Press; 1996. p. 507–40.
Haroche S, Raimond JM. Quantum computing: dream or nightmare? Physics Today. 1996;8:51–2.
Haynie D. Biological thermodynamics. Cambridge: Cambridge University Press; 2001.
Haynie DT. Biological thermodynamics. New York: Cambridge University Press; 2008.
Heisenberg W. Physics and philosophy. The revolution in modern science. New York: Harper and Row; 1958.
Hess B, Mikhailov A. Self-organization in living cells. Science. 1994;264:223–4.
Himelhoch S, Taylor SF, Goldman RS, Tandon R. Frontal lobe tasks, antipsychotic medication, and schizophrenic syndromes. Biol Psychiatry. 1996;39:227–9.
Hoffman RE, Kirstein L, Stopek S, Cicchetti DV. Apprehending schizophrenic dicourse: a structural analysis of the listener's task. Brain Lang. 1982;15:207–33.
Hofstadter DR. Gödel, Escher, Bach: an eternal golden braid. New York: Basic Books; 1999.
Igamberdiev AU. Physical limits of computation and emergence of life. Biosystems. 2007;90: 340–9.
Jain AK, Duin RPW, Mao J. Statistical pattern recognition: a review. IEEE Trans Pattern Anal Mach Intell. 2000;22:4–37.
Jensen O, Kaiser J, Lachaux JP. Human gamma-frequency oscillations associated with attention and memory. Trends Neurosci. 2007;30:317–24.
Jung CG. The psychology of dementia praecox. New York: Journal of Nervous and Mental Disease Publishing Company; 1909.(also in collected works of CG Jung 3).

Kantz H, Schreiber T. Nonlinear time series analysis. Cambridge: Cambridge University Press; 1997.

Kanwisher N. Neural events and perceptual awareness. Cognition. 2001;79:89–113.

Karolyhazy F. Gravitation and quantum mechanics of macroscopic objects. IL Nuovo Cimento A. 1966;42:309–42.

Kenardy J, Smith A, Spence SH, Lilley PR, Newcombe P, Dob R, et al. Dissociation in children's trauma narratives: an exploratory investigation. J Anxiety Disord. 2007;21:456–66.

Kent GH, Rosonoff AJ. A study of associations in insanity. Am J Insanity. 1910 66/67:37–34; 317–390.

Kline P. The handbook of psychological testing. New York, NY: Routledge; 2000.

Kragh H. Cosmology and controversy. Princeton: Princeton University Press; 1999.

Laurikainen KV. Beyond the atom: the philosophical thought of Wolfgang Pauli. Berlin: Springer Verlag; 1988.

Lee KH, Williams LM, Breakspear M, Gordon E. Synchronous gamma activity: a review and contribution to an integrative neuroscience model of schizophrenia. Brain Res Rev. 2003;41: 57–78.

Maniscalco S, Francica F, Zaffino RL, Lo Gullo N, Plastina F. Protecting entanglement via the quantum Zeno effect. Phys Rev Lett. 2008;100:090503.

Manschreck TC, Maher BA, Rucklos ME, White MT. The predictability of thought disordered speech in schizophrenic patients. Br J Psychiatry. 1979;134:595–601.

Manschreck TC, Maher BA, Ader DN. Formal thought disorder, the type token ratio and disturbed voluntary movement in schizophrenia. Br J Psychiatry. 1981;139:7–15.

Marshall IN. Consciousness and Bose-Einstein condensates. New Ideas Psychol. 1989;7:73–83.

Mashour GA. Consciousness unbound: toward a paradigm of general anesthesia. Anesthesiology. 2004;100:428–33.

Mashour GA. Toward a general theory of unconscious processes in psychoanalysis and anesthesiology. J Am Psychoanal Assoc. 2008;56:203–22.

Moran LJ, Mefferd Nr RB, Kimble Nr JP. Idiodynamic sets in word association. Psychol Monogr. 1964;78:1–22.

Mould RA. The inside observer in quantum mechanics. Found Phys. 1995;25:1621–9.

Mould RA. Consciousness and quantum mechanics. Found Phys. 1998;28:1703–18.

Nagel E, Newman J. Gödel's Proof. New York: New York Unversity Press; 2002.

Nielsen MA, Chuang IL. Quantum computation and quantum information. Cambridge: Cambridge University Press; 2000.

Pashler H. The psychology of attention. Cambridge MA: MIT Press; 1998.

Pauli W. Theory of relativity. New York: Dover Publications, Inc.; 1958.

Paulsen JS, Romero R, Chan A, Davis AV, Heaton RK, Jeste DV. Impairment of the semantic network in schizophrenia. Psychiatry Res. 1996;63:109–21.

Payne JD, Jackson ED, Ryan L, Hoscheidt S, Jacobs JW, Nadel L. The impact of stress on neutral and emotional aspects of episodic memory. Memory. 2006;14:1–16.

Peled A. Multiple constraint organization in the brain: a theory for schizophrenia. Brain Res Bull. 1999;49:245–50.

Penrose R. The Emperor´s new mind: concerning computers, minds, and the laws of physics. Oxford: Oxford University Press; 1989.

Penrose R. Shadows of the mind: an approach to the missing science of consciousness. Oxford: Oxford University Press; 1994.

Penrose R. The large the small and the human mind. Cambridge: The Press Syndicate of the University of Cambridge; 1997.

Penrose R. Consciousness, the brain, and spacetime geometry: an addendum. Ann N Y Acad Sci. 2001;929:105–10.

Penrose R. Gravitational collapse: the role of general relativity. Gen Relat Gravit. 2002;34: 1141–65.

Penrose R. The road to reality: a complete guide to the laws of the universe. London: Jonathan Cape; 2004.

Penrose R, Hameroff SR. What "gaps"? Reply to Grush and Churchland. J Conscious Stud. 1995;2:98–111.
Perrot P. A to Z of thermodynamics. New York: Oxford University Press; 1998.
Petitot J. The neurogeometry of pinwheels as a sub-Riemannian contact structure. J Physiol Paris. 2003;97:265–309.
Prigogine I, Stengers I. *Order out of chaos*: man's new dialogue with nature. New York: Bantam; 1984.
Rothstein J. Information, measurement, and quantum mechanics. Science. 1951;114:171–5.
Rucker R. Infinity and the mind: the science and philosophy of the infinite. Princeton, N.J.: Princeton University Press; 1982.
Sauseng P, Klimesch W. What does phase information of oscillatory brain activity tell us about cognitive processes? Neurosci Biobehav Rev. 2008;32:1001–13.
Schlosshauer M. Decoherence, the measurement problem, and interpretations of quantum mechanics. Rev Mod Phys. 2005;76:1267–305.
Schrödinger E. What is life? The physical aspect of the living cell. Cambridge: Cambridge University Press; 1944.
Schwartz JM, Stapp HP, Beauregard M. Quantum physics in neuroscience and psychology: a neurophysical model of mind-brain interaction. Philos Trans R Soc Lond B Biol Sci. 2005;360: 1309–27.
Scott A, editor. Encyclopedia of nonlinear science. New York: Routledge, Taylor and Francis Group; 2005.
Shakow D. Kent–Rosanoff association and its implications for segmental set theory. Schizophr Bull. 1980;6:676–85.
Shannon CE. A mathematical theory of communication. AT&T TECH J. 1948;27:379–423. and 623–656.
Shannon CE, Weaver W. A mathematical theory of communication. Champaign, IL: University of Illinois Press; 1963.
Smolin L. Atoms of space and time. Sci Am. 2004;290:66–75.
Sporns O, Tononi G, Edelman GM. Connectivity and complexity: the relationship between neuroanatomy and brain dynamics. Neural Netw. 2000;13:909–22.
Stapp HP. Quantum theory and the role of mind in nature. Found Phys. 2001;31:1465–99.
Stapp HP. Quantum leaps in philosophy of mind: reply to bourget's critique. J Conscious Stud. 2004;11:43–9.
Stapp HP. Quantum interactive dualism: an alternative to materialism. J Conscious Stud. 2005;12: 43–58.
Sudarshan ECG, Misra B. The Zeno's paradox in quantum theory. J Math Phys. 1977;18:756–63.
Szilard L. On the decrease of entropy in a thermodynamic system by the intervention of intelligent beings [In German published in 1929, Über die Entropieverminderung in einem thermodynamischen System bei Eingriffen intelligenter Wesen, *Zeitschrift für Physik, 53*, 840–856]. In English published in 1964, Behav Sci. 1929/1964;9:301–10.
Tegmark M. The importance of quantum decoherence in brain processes. Phys Rev E. 2000;61: 4194–206.
Teicher M, Andersen SL, Polcari A, Anderson CM, Navalta CP, Kim DM. The neurobiological consequences of early stress and childhood maltreatment. Neurosci Biobehav Rev. 2003; 27:3–44.
Teicher M, Tomoda A, Andersen SL. Neurobiological consequences of early stress and childhood maltreatment: Are results from human and animal studies comparable? Ann N Y Acad Sci. 2006;1071:313–23.
Tononi G, Edelman GM. Schizophrenia and the mechanisms of conscious integration. Brain Res Rev. 2000;31:391–400.
Tononi G, Edelman GM. Consciousness and complexity. Science. 1998;282:1846–51.
Tononi G, Edelman GM, Sporns O. Complexity and coherency: integrating information in the brain. Trends Cogn Sci. 1998a;2:474–84.

Tononi G, Sporns O, Edelman GM. A complexity measure for selective matching of signals by the brain. Proc Natl Acad Sci USA. 1996;93:3422–7.

Uhlhaas PJ, Singer W. Abnormal neural oscillations and synchrony in schizophrenia. Nat Rev Neurosci. 2010;11:100–13.

Uhlhaas PJ, Haenschel C, Nikolić D, Singer W. The role of oscillations and synchrony in cortical networks and their putative relevance for the pathophysiology of schizophrenia. Schizophr Bull. 2008;34:927–43.

Valentine ER. Neural nets: from Hartley and Hebb to Hinton. J Math Psychol. 1989;33:348–57.

Vinogradov S, Kirkland J, Poole JH, Drexler M, Ober BA, Shenaut GK. Both processing speed and semantic memory organization predict verbal fluency in schizophrenia. Schizophr Res. 2002;8:171–81.

Volkenstein MV. Entropy and information. Basel, Boston, Berlin: Birkhauser Verlag AG; 2009.

von Laue M. History of physics. New York: Academic Press; 1950.

Von Neumann J. Mathematical foundations of quantum mechanics. Princeton: Princeton University Press; 1955.

Wheeler JA, Zurek WH. Quantum theory and measurement. New Jersey: Princeton University Press; 1983.

Womelsdorf T, Fries P. Neuronal coherence during selective attentional processing and sensory-motor integration. J Physiol Paris. 2006;100:182–93.

Womelsdorf T, Fries P. The role of neuronal synchronization in selective attention. Curr Opin Neurobiol. 2007;17:154–60.

Woolf N, Hameroff S. A quantum approach to visual consciousness. Trends Cogn Neurosci. 2001;5:472–8.

Zaccai G, Massoulié J, David F. From cell to brain: the cytoskeleton, intra- and inter-cellular communication, the central nervous system. Amsterdam: Elsevier; 1998.

Zurek WH. Maxwell's Demon, Szilard's engine and quantum measurements. In: Moore GT, Scully MO, editors. Frontiers of nonequilibrium statistical physics. New York: Plenum Press; 1984. p. 151–60.

Chapter 7
The Universe Within

This chapter describes how these current findings may help to understand the relationship between the mental world and the physical world. There is evidence that causal connections exist between the mind and the body that point to their unitary nature. The causal influence of the mind enables it to manipulate the external world and influence its own body. Current findings indicate that the mental state may significantly influence the brain and body on various functional levels. These findings suggest a new paradigm for an understanding of the mental world and the possibility of influencing the physical body using psychotherapy. On the other hand, this interaction between mind and brain enables the laws of nature to be discovered and the external world to be understood through the rules that integrate the basic nature of the mind and the physical world.

It is said that the phrase "Let no one ignorant of geometry enter" guarded the door of Plato's Academy. In Plato's dialog *Meno* (Cooper 1997), Socrates asks a boy for geometrical questions about lines and squares that proceed to a higher level of abstraction, and then Socrates says to Meno: "Without anyone having taught him, and only through questions put to him, he will understand, recovering the knowledge out of himself …" (*Meno*, paragraph 85). Later, Socrates says to Meno: "And if the truth of all things that are is always in our soul, then the soul must be immortal …?" (*Meno*, paragraph 86).

Put simply, according to Plato, an ability to recognize and understand represents an innate part of the human mind, which, a priori, determines the ability to learn. If Socrates (or Plato) is right, it would mean that there are certain kinds of intuition related to preferred modalities of understanding and that there is the possibility of developing them through a kind of questioning. However, a key question is whether this ability to pre-understand is only an illusion or a part of our experience.

P. Bob, *Brain, Mind and Consciousness: Advances in Neuroscience Research*,

DOI 10.1007/978-1-4614-0436-1_7,

7.1 Mind and Mathematics

Although the question has been under debate since Plato's time (and earlier), its scientific study is relatively new and comes from the field of ethology, which was established as a specific discipline by Konrad Lorenz. Lorenz, who in his experimental research with geese, found that certain abilities to recognize are automatically activated without prior learning, and developed the concept of imprinting, which was proposed in the scientific literature for the first time in the 1870s by Douglas Spalding (Lorenz 1935/1970, 1950, 1965). Lorenz described these innate "fixed action patterns" as an underlying predisposition to learn with relatively stereotypical behavioral elements that are very similar in various populations of animals and can be analyzed in a similar way to anatomical and physiological properties (Lorenz 1950). Later neurobiological evidence has shown that these innate patterns of behavior are related to specific neural activities in multiple specialized memory systems that define a recognition pattern through various levels of binding (Freeman 1991, 2000; 2001; Fell et al. 2003; Womelsdorf and Fries 2006, 2007; Jensen et al. 2007; Sauseng and Klimesch 2008; Fries 2009). These specifically different patterns of neural activity enable distinction among various outside stimuli, for example, the distinction between memory for song and memory for spatial locations in birds, and similarly in humans who have specific memories of various specific behavioral patterns (Sherry and Schacter 1987; Goldsmith 1991; Sherry et al. 1992; Cosmides and Tooby 1995; Shettleworth 2009).

Put simply, it means that internally generated behavioral patterns are related to the specific ability to recognize, representing innate disposition, which is modified and developed by learning. However, the question is whether such a specific and complex ability of abstraction-like mathematics may be understood as an activation of specifically internally generated patterns of neural activity or whether it almost exclusively results from learning.

Current research into the comparative psychology of numerical cognition in animals and human infants is mainly focused on discrimination among numerosities (Brannon 2006; Shettleworth 2009). These data provide evidence that human infants and nonhuman primates have a rudimentary numerical ability that implies a relatively sophisticated representational system in which numbers are language-independent mental magnitudes, for example, in the absence of language in preverbal human infants who are capable of simple arithmetical operations, such as adding and subtracting a small number of visually presented objects (Pepperberg 1994; Hauser et al. 1996; Brannon 2006; Shettleworth 2009). Recent neuroscientific evidence indicates that numerical concepts have an ontogenetic origin and a neural basis that are independent of language (Gallistel and Gelman 1992; Gelman and Butterworth 2005).

Further behavioral and neuropsychological evidence indicates that these ontogenetically and phylogenetically shared abilities rest on the system for representing large and approximate numerical magnitudes, and on the system for the precise representation of small numbers of individual objects. However, although these systems constitute the basis for numerical intuitions and the ability to think in numbers preverbally, as is evident in human infants, primates and other animal species, further research has shown that representations underlying the ability to create abstract numerical concepts are present only in a subset of human adults (Feigenson et al. 2004). Taken together, recent neuroscientific research shows that mathematical intuition is an experimentally testable and valid phenomenon that is related to brain systems located in the intraparietal sulcus of both hemispheres and that neural representations of numerical information engage extensive cerebral networks (Nieder 2005; Dehaene 2009; Cantlon et al. 2009).

7.2 Savant Syndrome and Mathematical Intuition

Exceptional cases indicating intriguing information about mathematical intuition and the principles of its processing in cerebral networks provide findings in autistic individuals or in individuals with autistic spectrum disorders with the so-called savant syndrome. These cases document the outstanding mental capabilities of numerical processing and mathematical intuition that coexist with pervasive intellectual limitations and severe mental disabilities, such as deficits in social cognition, executive functions, self-involvement, and repetitive compulsive behavior (Treffert 1989, 1999, 2002, 2009). Although the usual mental abilities of these individuals are significantly reduced in many cases, they have exceptional abilities in arithmetic or calendrical calculations (Horwitz et al. 1965, 1969; O'Connor and Hermelin 1984, 1992; Hermelin and O'Connor 1986, 1990b; Howe, and Smith 1988; Hurst and Mulhall 1988; Ho et al. 1991; Mannheim et al. 1992; Rumsey et al. 1992; Moriarty et al. 1993; Nelson and Pribor 1993; Spitz 1995; Gonzalez-Garrido et al. 2002), musical skills (Viscott 1970; Hermelin et al. 1987, 1989; Miller 1987, 1989, 1995), and other special skills (Morishima and Brown 1977; Brink 1980; Abhyankar et al. 1981; Hermelin and O'Connor 1983, 1990a; O'Connor and Hermelin 1987a, b, 1989, 1990, 1991a, b; Howe 1989a; White 1988; Foerstl 1989; Patti and Lupinetti 1993) that are frequently linked to disturbed activity of the left cerebral hemisphere, leading to compensatory lateralization of cerebral functions in the right hemisphere (Delong 1999; Gazzaniga 2000; Corballis et al. 2000). A historical example of the special skills was documented by Rife and Snyder in 1931, when they reported a case of a man who was able to calculate in his mind the square root of four-digit numbers within 4 s and the cube root of six-digit numbers within 6 s (Rife and Snyder 1931; Hill 1978). Another interesting example of the special skills are savants who are able to quickly calculate the day of the week for a given date (calendrical calculations; for references see above). For example, Heavey et al. (1999) investigated eight savants with calendrical calculation skills and found that the savants did not differ from controls with regard to measures of general short- and long-term memory; nevertheless, they did show a clear superiority of recall for the long-term retention of calendrical material. Heavey et al. suggested that a general mnemonic advantage cannot explain the savants' date calculation skills and that the savants likely develop a structured knowledge utilizing calendar-related interrelations (Heavey et al. 1999).

Later epidemiological findings published by Rimland (1978) have shown that these special skills are not rare in autistic children. Rimland, using detailed questionnaires filled out by the parents of these children, studied the incidence of savant skills in 5,400 children and found that in 9.8% some special skills were reported, with a sex ratio of 3.54:1 (males:females).

Among these special skills related to mathematical intuition are the particularly interesting abilities of some savants to identify prime numbers that according to mathematical theory play an exceptional and central role in the arithmetic theory of numbers (Miller 1976; Bressoud 1989). An interesting case study of savants with an especially intriguing ability to identify prime numbers was reported by Oliver Sacks (1985), involving 26-year-old autistic twins John and Michael in 1966. The twins, during their game, spontaneously exchanged prime numbers, even though they did not know simple arithmetic and the task is very difficult even for people with exceptional mathematical skills. The prime numbers (or primes) are natural numbers (whole and positive), such as 2, 3, 5, 7, 11, 13, 17, 19, …, 71, 73, 79, 83, 89, 97, etc., that can be divided only by 1 or the number itself (without a remainder). To find them using mental calculation is a very difficult task, mainly for prime numbers with 3 or more digits. A simple method of identifying prime numbers was invented by the Greek philosopher Eratosthenes. Using this method, it is possible to find primes for numbers between 2 and *n*, where *n* can be, for example, 10. All the numbers in the interval are 2, 3, 4, 5, 6, 7, 8, 9, and 10. Then, simply delete all numbers that are divisible by two (2, 3, ~~4~~, 5, ~~6~~ 7, ~~8~~ 9 ~~10~~) and in the second step also delete all numbers divisible by three (2, 3, ~~4~~, 5, ~~6~~, 7, ~~8~~, ~~9~~, ~~10~~), and what remains after this procedure are the prime numbers (2, 3, 5, 7). This method works efficiently for the smaller primes below 10 million (10,000,000), i.e., for numbers with eight digits or less, but there is no simple method of calculation for primes of higher orders (Bressoud 1989). However, the twins, during their game with primes, at first created six-digit primes within seconds and then also larger, ten-digit prime numbers, which they calculated within 5 min. Later in the game they exceeded Sacks' tables of prime numbers.

Anderson et al. (1999) reported two experiments investigating a calculation strategy used by a low IQ savant to identify prime numbers using Eratosthenes' procedure and compared the reaction times that the savant needed to decide whether a number was prime or not and then the reaction times were compared with a control subject proficient in mathematical calculation. Results have shown that the reaction times of the savant were faster and that his speed of processing was far superior to that expected from someone of his IQ. Interesting cases of savants able to identify prime numbers were also reported by Welling (1994) and Anderson et al. (1999). Welling found that many patients who demonstrate this ability do not possess any of the arithmetical skills needed to perform such calculations. He has also suggested that a distinction between prime and nonprime numbers might be made if there is an ability to recognize the primes through a spontaneous tendency to find symmetry (Welling 1994).

An intriguing example of how this tendency and ability to find symmetry might be performed was demonstrated by autistic savant Daniel Tammet, who is well-known because of his European record for the number of digits of pi that he was able to know from memory (i.e., 22,514 digits) (Biever 2009). However, in contrast to many other savants, he was able to describe how he did it. Tammet reported his subjective experiences of seeing numbers in spatial forms with sizes, colors, and texture where prime numbers have special properties (like color, position, etc.) that distinguish them from other numbers (Tammet 2006; Baron-Cohen et al. 2007; Murray 2010). Based on similar data, Azouli et al. (2005) proposed synesthesia as a possible mechanism of savant skills. Together, these data suggest that the ability to recognize prime numbers might represent intuitive intelligence, reflecting the mathematically exceptional and fundamental role of the prime numbers and their sequences in mathematics and physics (for example, see Miller 1976; Agrawal et al. 2004; Iovane et al. 2004; Iovane 2008).

Snyder (2009) proposed an interesting hypothesis about the neurophysiological nature of savant skills. Based on research data focused on partial inactivation of the left hemisphere using low-frequency repetitive transcranial magnetic stimulation (rTMS) in healthy people, he suggests that such skills might be artificially induced. In agreement with experimental data, Snyder (2009) suggests that using low-frequency rTMS might temporarily inhibit neural activity in a localized area of the cerebral cortex, enabling "virtual lesions" to be created (Hilgetag et al. 1999; Walsh and Cowey 2000; Hoffman and Cavus 2002) and top–down inhibitory influences to be decreased, which may enable access to the lower levels of information processes that savants have (Snyder 2009).

In this context, Snyder (2009) suggests that savant-like skills might also be artificially induced in normal healthy individuals by inhibiting the left anterior temporal lobe, in agreement with data that autistic savants have left brain dysfunction and atypical inhibitory functions of the left hemisphere (Wilson et al. 2007) that are compensated for by the right hemisphere (Miller et al. 1998; Treffert 2005; Sacks 2007), which leads to a predilection for literal, nonsymbolic skills (Sacks 2007, pp. 314–315; Treffert 2005, 1989). This explanation is also in agreement with the right-hemispheric bias that is frequently associated with autism (Herbert et al. 2005; Koshino et al. 2005). In support of this hypothesis, Snyder et al. (2003) have found significant changes in drawing style following active rTMS stimulation for 15 min over the left anterior temporal lobe in 11 right-handed, healthy participants. In this experiment the participants were given 1 min to draw a dog, horse or face from memory for a time before, during, immediately after, and 45 min after rTMS stimulation. They found that the stimulation caused major changes in the drawing style in 4 of the 11 participants, whose drawings returned to their normal style 45 min after rTMS stimulation. The changes in drawing were characterized by a heightened focus on seemingly meaningless details and a naturalistic drawing style that may be explained by heightened access to raw, less-processed sensory information typical of savants' performance (Snyder and Mitchell 1999; Snyder 2009). This may be caused by sensory hypersensitivity and enhanced perception of details (Minshew and Hobson 2008), which likely exists, not only in savants, but also in normal individuals, in whom the information is inaccessible because of top–down inhibition (Snyder et al. 2003). Similar data were reported by Young et al. (2004), who, using repetitive transcranial magnetic stimulation of the frontotemporal lobe, found that savant-like skills improved in 5 out of 17 participants during the time of stimulation. Those participants with savant-like skills reported greater awareness of detail in their surroundings and enhanced skills, which included declarative memory, drawing, mathematics, and calendar calculating.

Similar savant-like skills caused by interruption of the frontotemporal lobe functions were also reported by Miller et al. (1996, 1998). In their 1996 study, Miller et al. interviewed 69 frontotemporal dementia patients regarding their visual abilities and found 5 patients who had become artists and acquired new artistic skills in the early stages of dementia. They found that 4 of the 5 patients with the temporal variant of frontotemporal dementia involving the anterior temporal lobes had preserved functions of the dorsolateral frontal cortex. These 4 patients had preserved visual skills, but their language and social skills were impaired, which suggests that the loss of function in the anterior temporal lobes might influence facilitation of artistic skills. Similar and very interesting case studies were reported by the same authors in their 1998 study. They documented three previously nonartistic patients who developed skills in art later in life, related to their progressive dementia, and became accomplished painters. All three patients had a temporal variant of frontotemporal dementia with dysfunctional anterior temporal lobes and relatively spared frontal lobes, and they improved artistically during the early and middle stages of their illness. Together, these clinical observations suggest that the brain dysfunction restricted to the anterior temporal lobe may be associated with enhanced artistic ability, which may be explained by dysfunction of the anterior temporal lobe systems, leading to decreased inhibitory functions of these structures. This inhibitory failure influences usual contexts of information processing, leading to deficits in attentional filtering and less filtered information, including visual experiences and memories, and an unusual sensitivity to detail (Miller et al. 1998).

This sensitivity to detail related to the inhibitory failure may explain the unusual ability of some autistic savants to process mathematical information in a way that indicates mathematical intuition, but without the normal ability to integrate these spontaneous insights into concepts that enable fragmentary data to be integrated in mutual connections. This mode of mental processing likely also reflects specific changes in brain information processing, which, according to recent evidence, is less globally integrated with long-distance under-connectivity and with a higher level of integration on some local levels, leading to local over-connectivity (Wass 2011) and hypersynchronization, which may influence vulnerability to epilepsy (Nomura et al. 2010; Parmeggiani et al. 2010; Tuchman et al. 2010). This predominant method of information processing in autistic patients likely indicates a lower level of binding and deficits in large-scale information integration in the brain. In this context, focus on detail and deficits in the ability to integrate new information from various sources likely related to deficits in global brain integration and binding may help to explain why autistic individuals are not able to connect various modes of information in mutual connections and create adaptive conscious awareness that enables understanding of novel contexts and integration of new information into the coherent whole. On the other hand, these specific cognitive changes in autistic individuals that enable unusual mathematical skills constitute an important topic for further research that could help to explain some of the basic neurocognitive mechanisms of mathematical intuition and human intelligence.

7.3 Mathematical Intuition and the Universe Within

Although some autistic savants demonstrate unusual examples of mathematical intuition and likely have insights into mathematical symmetry, a typical deficit of these individuals in their processing of mathematical information is the inability to integrate fragmentary insights into mutual connections and create abstract numerical and mathematical concepts. The intriguing ability of the savants to process details is likely based on changes in brain mechanisms that enable multilevel information processing regulated within the framework of contextual understanding, which strictly determines what details are important for the whole coherent context. For this purpose, the brain has several areas in the frontal and temporal regions that are on the upper level of the hierarchical processing, which may provide feedback to lower levels and may initiate reprocessing of any information that enables its reintegration in the contexts of various schemes and concepts (Mesulam 1998; Nadel and Jacobs 1998; Lavenex and Amaral 2000). For example, in visual research this hierarchical processing has been found in several frontal and temporal regions that may provide inhibitory feedback to the initial processing in the primary visual areas, V1 and V2 (Hupe et al. 1998; Bullier 2006), which likely has a specific deficit in autism (Fabricius 2010). This hierarchical connectivity of the brain areas that enables the information to be reprocessed within contextual frameworks fails in the autistic brain, in which detailed reprocessing does not occur within a new contextual framework, but is predominantly linked only to heightened focus on the detailed structure of the information.

These findings suggest that the multilevel processing of perceptual information might be related to executive signals that in the framework of the required context determine sensitivity to various details needed for contextual processing. For example, in visual processing several data suggest that the laminar circuits of the visual cortical areas, V1 and V2, implement context-sensitive binding processes that may significantly influence processing of visual scenes through various feedback interactions (Gilbert and Wiesel 1990; Francis et al. 1994; Gove et al. 1995; Gray 1999; Grossberg and Grunewald 1997; Ito and Gilbert 1999; Raizada and Grossberg 2001). Thus, contextual processing as an ability to selectively attend and respond to environmental events relies on the brain's ability to self-regulate its arousal and attentional focus during information processing (Kang et al. 2005).

In this sense, contextual processing may determine what details of the processed information will actually be presented in conscious awareness, while other detailed information related to this percept may potentially be processed and included in other contextual frameworks, which in usual language is included in the statement: "The one truth has many sides" (or as Socrates said to Meno: "... only through questions ... recovering the knowledge out of himself ..." *Meno*, paragraph 85). Consequently, these contextual mechanisms of attentional sensitivity may influence the processing of various aspects of perceptual information and enable events to be discovered that in certain contextual frameworks are presented as hidden (subliminal) parts of the human conscious experience.

Connectivity patterns related to specific states of consciousness and attentional sensitivity are mediated through various levels of binding and large-scale brain information processing that enable information from various sensory modalities to be combined with information in memory and create coherent contextual frameworks (Fries 2005; Singer 2001; Varela et al. 2001; Zeman 2001; John 2002; Crick and Koch 2003; Bob 2009). Through these various modes of binding in brain information processing, it is possible not only to create various contextual schemes, but also to experience various details of processed information that are possible to realize only within a specific cognitive scheme. In this context, dissociation presents a mental event in which some contextual element is conflicting, and although it is a part of the mental scheme, at the same time, it does not fit into this contextual framework. This process is subjectively experienced as conflicting and provokes a need to find internal integrity and symmetry, which is possible using various methods of psychotherapy that enable a new cognitive scheme to be found, providing the possibility of integrating a dissociated memory of the event and performing memory reconsolidation, which significantly influences the brain and body on various functional levels (Nadel and Jacobs 1998; Payne et al. 2006; Bob 2007). These findings suggest a new paradigm for understanding the mental world and the possibility of influencing the physical body using psychotherapy through specific changes in contextual mental processing related to the learning process that may specifically influence and modify brain functions, metabolism in specific brain structures, and also genetic processes (Kandel 1998, 1999; Gabbard 2000, 2007).

On the other hand, this interaction between mind and brain enables one to discover the laws of nature and understand the external world through the rules that integrate the basic nature of the mind and the physical world. These ideas suggest a specific role for attentional sensitivity in the process of understanding the world that, through more detailed introspective experience and increased sensitivity of measurement, enables scientific description of processes in the external material world.

An intriguing relationship between descriptions of the internal and external world occurs in mathematical theory. Although formation of mathematical concepts and theories is a mental event, and mathematics as such is not a scientific discipline, it is possible to use mathematical language for the description of various processes in the physical world. In addition, some important mathematical theories were established without direct connection to experimental science and their discovery was based mainly on mathematical intuition and abstraction. Exceptional examples of such mathematical theories in the history of science are differential, and the integral calculus that was the mathematical basis for the Newtonian formulation of

Fig. 7.1 Isaac Newton

mechanics, and a few centuries later developed Riemannian geometry, constitutes the mathematical language for Einstein's general theory of relativity.

Radical new discoveries in mathematics and the physical sciences in the second half of the seventeenth century were made by Isaac Newton (Fig. 7.1), who (at the same time and independently of G.W. Leibniz) developed an analytical (differential and integral) calculus and applied it to the theory of mechanics. Newton later also used this calculus as a basis for his concept of a universal gravitational force and found a beautiful and simple analytical formulation that framed his philosophical ideas concerning the gravitational force and its role in the universe (Brackenridge 1995). The basic principle of this calculus, applied to natural philosophy, is that infinitely small entities, when they are put together, may create finite entities like objects or processes in the universe. In this sense, very small entities like atoms and other elementary particles that may be infinitely small when, for example, they reach Planck's length ($\approx 10^{-35}$ m; see Chap. 6.2.), may enable creation of the universe. What is paradoxical and extremely intriguing about the differential and integral analytical calculus is that the "multiplication" of infinitely small size may provide a finite result, for example, the summation of infinitely small volumes may provide a finite volume of a body, or a summation of infinitely small distances may provide a finite distance between places. From this point of view, infinitely small size exists, although, for example, no division of macroscopic volumes reaches this infinitesimally small entity. Practical application of this calculus in physics enables very intriguing findings and technological procedures. This brilliant and intuitively abstract mathematical concept of "infinitely small," although it is paradoxical with respect to our everyday experience, represents part of the universe.

Newton developed the problem of infinitely small size in his *Tract on Fluxions*, written in 1666 (Whiteside 1967), where he proposed a basic formal structure of the analytical differential and integral calculus that he later applied to his physical study in the historically groundbreaking work *Philosophia Naturalis Principia Mathematica.*

In his Preface to the first edition of his *Principia* (*Philosophia Naturalis Principia Mathematica*, translated into English as *Mathematical Principles of Natural Philosophy*), Newton (1686/1974, p. XVIII) wrote:

> I derive from the celestial phenomena the forces of gravity with which bodies tend to the sun and the several planets. Then from these forces, by other propositions which are also mathematical, I deduce the motions of the planets, the comets, the moon, and the sea. I wish we could derive the rest of the phenomena of Nature by the same kind of reasoning from mechanical principles, for I am induced by many reasons to suspect that they may all depend upon certain forces by which the particles of bodies, by some causes hitherto unknown, are either mutually impelled towards one another, and cohere in regular figures, or are repelled and recede from one another. These forces being unknown, philosophers have hitherto attempted the search of Nature in vain; but I hope the principles here laid down will afford some light either to this or some truer method of philosophy ....

In this intriguing context, Newton developed his brilliant work and used novel mathematics as a tool and part of the universal mechanics, and discovered the science of motions resulting from forces in the universe that connect heaven and earth within the same rules.

A few centuries later, in 1854, Bernhard Riemann presented his lecture on the geometry of higher dimensions and developed other intriguing mathematical ideas, for example, his well-known theorem about prime numbers, known as "Riemann's hypothesis" (Monastyrsky 1999; Derbyshire 2004). In 1868, Riemann published his seminal work *Über die Hypothesen welche der Geometrie zu Grunde liegen* (*On the Hypotheses Which Underlie Geometry*), in which he developed the basic concept of Riemannian geometry, which, as already mentioned above, was the basis for the mathematical language for Einstein's general theory of relativity (Monastyrsky 1999; Plotnitsky 2009). The most important application of Riemann's ideas was developed for the mathematical description of space–time that is a four-dimensional space, consisting of three spatial dimensions plus time. Riemann found that specific changes in space may be described by a collection of numbers that sufficiently characterize the properties of every point in this space, and enables one to characterize how curved the space is. The level of curvature characterizes a difference between the usual "flat" Euclidean space and the curved space with Riemannian metrics (Monastyrsky 1999; Plotnitsky 2009). An intriguing and groundbreaking idea about how to apply Riemannian metrics was published by Einstein (Fig. 7.2) in 1915, who found that the origin of gravitation described by Newton as a gravitational force may be better described as a specific change in spatial structure that may be explained as geometrical curvature, which changes space–time metrics (Einstein 1916/2005; Laue 1950; Pauli 1958; Beiser 1995).

Einstein's formulation of the general theory of relativity as a theory of gravitation later led to numerous physical applications in the field of cosmology. For example, Einstein's theory directly implies the existence of black holes, which represent regions in space with extremely curved Riemannian metrics, leading to an extreme distortion of space and time and an accumulation of extreme mass, energy, and gravitation (Thorne 1994; Carmeli 1999; Field 1999). Another intriguing finding that enables the general theory of relativity is the estimation of the total amount of matter in the universe.

Fig. 7.2 Albert Einstein

These findings suggest that only the smaller amount of matter in the universe is the "usual" matter that constitutes objects like stars, planets, and other space objects that have observable physical characteristics and interact with the electromagnetic field and its radiation, like light. The majority of matter in the universe appears to be the so-called dark matter, which has only mass and gravitation, likely does not have other known physical characteristics of matter, and does not interact with an electromagnetic field and (similar to black holes) cannot be directly observed (Carmeli 1999; Capozziello and Sarkar 2005; Capozziello and Francaviglia 2008; Zakharov et al. 2009; Bertone 2010).

What is intriguing about mental processing related to mathematics is an ability to create an abstract experience of rules, symmetry, and structural order that cannot be simply deduced from sensory experience. For example, without Riemannian geometry, which was discovered about 50 years before the general theory of relativity, it would likely not be possible to know anything about the black holes and other curious objects and experimental predictions that are significantly different from the usually expected behavior described within the framework of Newtonian mechanics. In addition, experimental testing of any novel theory is frequently a difficult task and Einstein repeatedly faced attacks because of his scientific works. In this context, evidence that the abstract scheme, symmetry, and structural order discovered in the human mind may so profoundly explain the physical world is a mystery of the mind, particularly in cases when the mathematical theory is developed relatively independently of its physical applications and its hypothetical predictions contradict usual everyday or actual scientific experience. In this sense, reading mathematics in the human mind likely presents a reflection of its own rules at a deeper level of its functioning that are intriguingly similar to the rules that occur in the external world, which enables us to use this formal mental structure for a scientific understanding of the universe, as Newton, Einstein, and others have shown.

On the other hand, mind-"reading" mathematics is extremely focused on its internal universe, which implies that it may be autistic-like, simply because it is not possible to read the Bible and daily news or journals at the same time. This contemplative and autistic-like nature of mental processing related to mathematics may explain why some great mathematicians and physicists like Newton, Einstein, and others were thought to have a mild form of autism or Asperger syndrome (Asperger 1991; James 2003; Hazel 2003; Fitzgerald 2004).

However, independently of what we know about the nature of mathematical knowledge, the interaction between the mind and brain enables us to discover the laws of nature and understand the external world through the rules that connect observed objects and processes. It likely also enables the basic nature of the mind and the physical world to be integrated. As Socrates said to Meno: "And if the truth of all things that are is always in our soul, then the soul must be immortal …?" (*Meno*, paragraph 86).

References

Abhyankar R, Thatte S, Doongaji D. Idiot *savant*. J Postgrad Med. 1981;22:44–7.
Agrawal M, Kayal N, Saxena N. PRIMES is in P. Ann Math. 2004;160:781–93.
Anderson M, O'Connor N, Hermelin B. A specific calculating ability. Intelligence. 1999;26: 383–403.
Asperger H. Autistic psychopathy in childhood (U. Frith, Trans.). In Frith U, editor. Autism and asperger syndrome. Cambridge, England: Cambridge University Press; 1991. p. 37–92. (Original work published 1944).
Azouli S, Hubbard EM, Ramachandran VS. Does synesthesia contribute to mathematical savant skills? J Cognitive Neurosci. 2005; 69, 1 (Supl). http://psy2.ucsd.edu/~edhubbard/abstracts/Azoulai_CNS05.pdf.
Baron-Cohen S, Bor D, Billington J, Asher J, Wheelwright S, Ashwin C. Savant memory in a man with colour form-number synaesthesia and asperger syndrome. Journal of Consciousness Studies. 2007;14:237–51.
Beiser A. Concepts of modern physics. 5th ed. New York: Mc Graw Hill; 1995.
Bertone G. The moment of truth for WIMP dark matter. Nature. 2010;468:389–93.
Biever C. Peek inside a singular mind. New Scientist, 2689; 2009:40–41. Online version: Inside the mind of an autistic savant. New Scientist. http://www.newscientist.com/article/mg20126881.800-inside-the-mind-of-an-autistic-savant.html.
Bob P. Hypnotic abreaction releases chaotic patterns of electrodermal activity during dissociation. Int J Clin Exp Hypn. 2007;55:435–56.
Bob P. Quantum science and the nature of mind. J Mind Behav. 2009;30:1–14.
Brackenridge JB. The key to Newton's. dynamics: the Kepler problem and the Principia. Berkeley, CA: University of California Press; 1995.
Brannon EM. The representation of numerical magnitude. Curr Opin Neurobiol. 2006;16:222–9.
Bressoud DM. Factorizations and primality testing. New York: Springer-Verlag; 1989.
Brink T. Idiot *savant* with unusual mechanical ability: an organic explanation. Am J Psychiatry. 1980;137:250–1.

Bullier J. What is fed back? In: van Hemmen J, Senjowski T, editors. Problems in systems neuroscience. New York: Oxford University Press; 2006. p. 103–32.
Cantlon JF, Platt ML, Brannon EM. Beyond the number domain. Trends Cogn Sci. 2009;13: 83–91.
Capozziello S, Francaviglia M. Extended theories of gravity and their cosmological and astrophysical applications. Gen Relat Gravit. 2008;40:357–420.
Carmeli M. Aspects of cosmological relativity. In J Theor Phys. 1999;38:1993–2007.
Cooper JM, editor. Plato: complete works. Indianapolis: Hackett; 1997. [Web source: Online Library of Liberty; Plato, The Dialogues of Plato, in 5 vols (Jowett editor.)] http://oll.liberty fund.org/?option=com_staticxtandstaticfile=show.php%3Ftitle=166andItemid=27.
Corballis PM, Funnell MG, Gazzaniga MS. An evolutionary perspective on hemispheric asymmetries. Brain Cogn. 2000;43:112–7.
Cosmides L, Tooby J. From function to structure: the role of evolutionary biology and computational theories in cognitive neuroscience. In: Gazzaniga M, editor. The cognitive neurosciences. Cambridge, MA: MIT Press; 1995. p. 1199–210.
Crick F, Koch C. A framework for consciousness. Nat Neurosci. 2003;6:119–26.
Dehaene S. Origins of mathematical intuitions: the case of arithmetic. Ann N Y Acad Sci. 2009;1156:232–59.
Delong RG. Autism: new data suggest a new hypothesis. Neurology. 1999;52:911–6.
Derbyshire J. Prime obsession: Bernhard Riemann and the greatest unsolved problem in mathematics. New York: Penguin; 2004.
Einstein A. Relativity the special and general theory. Pi Press: New York (Originally published 1916); 1916/2005.
Fabricius T. The savant hypothesis: is autism a signal-processing problem? Med Hypotheses. 2010;75:257–65.
Feigenson L, Dehaene S, Spelke E. Core systems of number. Trends Cogn Sci. 2004;8:307–14.
Fell J, Fernández G, Klaver P, Elger CE, Fries P. Is synchronized neuronal gamma activity relevant for selective attention? Brain Res Rev. 2003;42:265–72.
Field G. Astrophysics. Rev Mod Phys. 1999;71 Suppl 2:S33–40.
Fitzgerald M. Autism and creativity: is there a link between autism in men and exceptional ability? East Sussex: Brunner-Routledge; 2004.
Foerstl J. Early interest in *the* idiot *savant*. Am J Psychiatry. 1989;146:566.
Francis G, Grossberg S, Mingolla E. Cortical dynamics of feature binding and reset: control of visual persistence. Vision Res. 1994;34:1089–104.
Freeman WJ. The physiology of perception. Sci Am. 1991;264:78–85.
Freeman WJ. Mesoscopics neurodynamics: from neuron to brain. J Physiol Paris. 2000;94: 303–22.
Freeman WJ. Biocomplexity: adaptive behavior in complex stochastic dynamical systems. Biosystems. 2001;59:109–23.
Fries P. A mechanism for cognitive dynamics: neuronal communication through neuronal coherence. Trends Cogn Sci. 2005;9:474–80.
Fries P. Neuronal gamma-band synchronization as a fundamental process in cortical computation. Annu Rev Neurosci. 2009;32:209–24.
Gabbard GO. A neurobiologically informed perspective on psychotherapy. Br J Psychiatry. 2000;177:117–22.
Gabbard GO. Psychotherapy in psychiatry. Int Rev Psychiatry. 2007;19:5–12.
Gallistel CR, Gelman R. Preverbal and verbal counting and computation. Cognition. 1992;44: 43–74.
Gazzaniga MS. Cerebral specialization and interhemispheric communication: does the corpus callosum enable the human condition? Brain. 2000;123:1293–326.
Gelman R, Butterworth B. Number and language: how are they related? Trends Cogn Sci. 2005; 9:6–10.
Gilbert CD, Wiesel TN. The influence of contextual stimuli on the orientation selectivity of cells in primary visual cortex of the cat. Vision Res. 1990;30:1689–701.

Goldsmith TH. The biological roots of human nature: forging links between evolution and behavior. New York: Oxford University Press; 1991.
Gonzalez-Garrido AA, Ruiz-Sandoval JL, Gomez-Velazquez FR, de Alba JL, Villasenor-Cabrera T. Hypercalculia in savant syndrome: central executive failure? Arch Med Res. 2002;33: 586–9.
Gove A, Grossberg S, Mingolla E. Brightness perception, illusory contours, and corticogeniculate feedback. Vis Neurosci. 1995;12:1027–52.
Gray CM. The temporal correlation hypothesis of visual feature integration: still alive and well. Neuron. 1999;24:31–47.
Grossberg S, Grunewald A. Cortical synchronization and perceptual framing. J Cogn Neurosci. 1997;9:117–32.
Hauser MD, MacNeilage P, Ware M. Numerical representations in primates. Proc Natl Acad Sci USA. 1996;93:1514–7.
Hazel M. Einstein and Newton showed signs of autism. New Scientist. 2003. http://www.newscientist.com/article.ns?id=dn3676.
Heavey L, Pring L, Hermelín B. A date to remember: the nature of memory in savant calendrical calculators. Psychol Med. 1999;29:45–60.
Herbert MR, Ziegler DA, Deutsch CK, O'Brien LM, Kennedy DN, Filipek PA, et al. Brain asymmetries in autism and developmental language disorder: a nested whole-brain analysis. Brain. 2005;128:213–26.
Hermelin B, O'Connor N. *The* idiot *savant*: flawed genius or clever Hans? Psychol Med. 1983;13:479–81.
Hermelin B, O'Connor N. Idiot *savant* calendrical calculators: rules and regularities. Psychol Med. 1986;16:885–93.
Hermelin B, O'Connor N. Art and accuracy: *the* drawing ability of idiot-savants. J Child Psychol Psychiatry. 1990a;31:217–28.
Hermelin B, O'Connor N. Factors and primes: a specific numerical ability. Psychol Med. 1990b;20:163–9.
Hermelin B, O'Connor N, Lee S. Musical inventiveness of five idiot savants. Psychol Med. 1987;17:685–94.
Hermelin B, O'Connor N, Lee S, Treffert D. Intelligence and musical improvisation. Psychol Med. 1989;19:447–57.
Hilgetag CC, Kotter R, Young MP. Inter-hemispheric competition of sub-cortical structures is a crucial mechanism in paradoxical lesion effects and spatial neglect. Prog Brain Res. 1999;121:121–41.
Hill AL. Mentally retarded individuals with special skills. In: Eller NR, editor. International review of research in mental retardation, vol. 9. NewYork: Academic Pan; 1978. p. 277–98.
Ho E, Tsang A, Ho D. An investigation of *the* calendar calculation ability of a Chinese calendar *savant*. J Autism Dev Disord. 1991;21:315–27.
Hoffman RE, Cavus I. Slow transcranial magnetic stimulation, long-term depotentiation, and brain hyperexcitability disorders. Am J Psychiatry. 2002;159:1093–102.
Horwitz W, Kestenbaum C, Person E, Jarvich L. Identical twin "idiots savants" calendar calculators. Am J Psychiatry. 1965;121:1075–9.
Horwitz W, Deming W, Winter R. A further account of *the* idiots savants, experts with *the* calendar. Am J Psychiatry. 1969;126:412–5.
Howe M. Fragments of genius: the strange feats of idiot savants. London: Routledge; 1989.
Howe M, Smith J. Calendar calculating in "idiots savants:" how do they do it? Br J Psychol. 1988;79:371–86.
Hupe JM, James AC, Payne BR, Lomber SG, Girard P, Bullier J. Cortical feedback improves discrimination between figure and background by V1, V2 and V3 neurons. Nature. 1998;394:784–7.
Hurst L, Mulhall D. Another calendar *savant*. Br J Psychiatry. 1988;152:274–7.
Iovane G. The set of prime numbers: symmetries and supersymmetries of selection rules and asymptotic behaviours. Chaos, Solitons and Fractals. 2008;37:950–61.
Iovane G, Laserra E, Tortoriello FS. Stochastic self-similar and fractal universe. Chaos, Solitons and Fractals. 2004;20:415–26.

Ito M, Gilbert CD. Attention modulates contextual influences in the primary visual cortex of alert monkeys. Neuron. 1999;22:593–604.
James I. Singular scientists. J R Soc Med. 2003;96:36–9.
Jensen O, Kaiser J, Lachaux JP. Human gamma-frequency oscillations associated with attention and memory. Trends Neurosci. 2007;30:317–24.
John ER. The neurophysics of consciousness. Brain Res Rev. 2002;39:1–28.
Kandel ER. A new intellectual framework for psychiatry. Am J Psychiatry. 1998;155:457–69.
Kandel ER. Biology and the future of psychoanalysis: a new intellectual framework for psychiatry revisited. Am J Psychiatry. 1999;156:505–24.
Kang K, Williams LM, Hermens D, Gordon E. Neurophysiological markers of contextual processing: the relationship between P3b and gamma synchrony and their modulation by arousal, performance and individual differences. Brain Res Cogn Brain Res. 2005;25:472–83.
Koshino H, Carpenter PA, Minshew NJ, Cherkassky VL, Keller TA, Just MA. Functional connectivity in an fMRI working memory task in high-functioning autism. Neuroimage. 2005;24:810–21.
Lavenex P, Amaral DG. Hippocampal-neocortical interaction: a hierarchy of associativity. Hippocampus. 2000;10:420–30.
Lorenz K. Companions as factors in the bird's environment. In: Martin RD, editor. Studies in animal and human behavior, vol. 1. London: Methuen, and Co; 1935/1970. p. 101–258.
Lorenz K. The comparative method in studying innate behaviour patterns. Symp Soc Exp Biol. 1950;4:221–68.
Lorenz K. Evolution and modification of behavior. Chicago: University of Chicago Press; 1965.
Mannheim G, Krasnewich D, Rumsey J, Hibbs T. Calendar calculating savants (CCS): a distinct phenotype/phenomenology from autism. Neurology. 1992;42 Suppl 3:462.
Mesulam MM. From sensation to cognition. Brain. 1998;121:1013–52.
Miller GL. Riemann's hypothesis and tests for primality. J Comput Sys Sci. 1976;13:300–17.
Miller LK. Sensitivity to tonal structure in a developmentally disabled musical *savant*. Am J Ment Defic. 1987;91:467–71.
Miller LK. Musical savants: exceptional skill in the mentally retarded. Hillsdale, NJ: Erlbaum; 1989.
Miller LK. Sensitivity to sequential structure in musical savants. Am J Ment Retard. 1995;99:391–9.
Miller BL, Ponton M, Benson DF, Cummings JL, Mena I. Enhanced artistic creativity with temporal lobe degeneration. Lancet. 1996;348:1744–5.
Miller BL, Cummings J, Mishkin F, Boone K, Prince F, Ponton M, et al. Emergence of artistic talent in frontotemporal dementia. Neurology. 1998;51:978–82.
Minshew NJ, Hobson JA. Sensory sensitivities and performance on sensory perceptual tasks in high-functioning individuals with autism. J Autism Dev Disord. 2008;38:1485–98.
Monastyrsky M. Riemann, topology and physics. Boston, MA: Birkhauser; 1999.
Moriarty J, Ring H, Robertson M. An idiot *savant* calendar calculator with Gilles de la Tourette *syndrome*: implications for an understanding of *savant syndrome*. Psychol Med. 1993;23:1019–21.
Morishima A, Brown L. A case report of *the* artistic talent of an autistic idiot *savant*. Ment Retard. 1977;15:33–7.
Murray AL. Can the existence of highly accessible concrete representations explain savant skills? Some insights from synaesthesia. Med Hypotheses. 2010;74:1006–12.
Nadel L, Jacobs WJ. Traumatic memory is special. Curr Dir Psychol Sci. 1998;7:154–7.
Nelson E, Pribor E. A calendar *savant* with autism and Tourette *syndrome*: response to treatment and thoughts on *the* interrelationships of these conditions. Ann Clin Psychiatry. 1993;5:135–40.
Newton I. Mathematical principles of natural philosophy. Berkeley: University of California Press; 1686/1974.
Nieder A. Counting on neurons: the neurobiology of numerical competence. Nat Rev Neurosci. 2005;6:177–90.
Nomura Y, Nagao Y, Kimura K, Hachimori K, Segawa M. Epilepsy in autism: a pathophysiological consideration. Brain Dev. 2010;32:799–804.

O'Connor N, Hermelin B. Idiot *savant* calendrical calculators: math or memory? Psychol Med. 1984;14:801–6.
O'Connor N, Hermelin B. Visual and graphic abilities of *the* idiot *savant* artist. Psychol Med. 1987a;17:79–90.
O'Connor N, Hermelin B. Visual memory and motor programmes: their use by idiot-*savant* artists and controls. Br J Psychol. 1987b;78:307–23.
O'Connor N, Hermelin B. *The* memory structure of autistic idiot *savant* mnemonists. Br J Psychol. 1989;80:97–111.
O'Connor N, Hermelin B. *The* recognition failure and graphic success of idiot-*savant* artists. Journal of Child Psychology and Psychiatry. 1990;31:203–15.
O'Connor N, Hermelin B. A specific linguistic ability. Am J Ment Retard. 1991a;95:673–81.
O'Connor N, Hermelin B. Talents and pre-occupations in idiots savants. Psychol Med. 1991b;21:959–64.
O'Connor N, Hermelin, B. Do young calendrical calculators improve with age? J Child Psychol Psychiatry. 1992;33:907–12.
Parmeggiani A, Barcia G, Posar A, Raimondi E, Santucci M, Scaduto MC. Epilepsy and EEG paroxysmal abnormalities in autism spectrum disorders. Brain Dev. 2010;32:783–9.
Patti P, Lupinetti L. Brief report: implications of hyperlexia in an autistic *savant.* J Autism Dev Disord. 1993;23:397–405.
Pauli W. Theory of relativity. New York: Dover Publications, Inc.; 1958.
Payne JD, Jackson ED, Ryan L, Hoscheidt S, Jacobs JW, Nadel L. The impact of stress on neutral and emotional aspects of episodic memory. Memory. 2006;14:1–16.
Pepperberg IM. Numerical competence in an African gray parrot (Psittacus erithacus). J Comp Psychol. 1994;108:36–44.
Plotnitsky A. Bernhard Riemann's conceptual mathematics and the idea of space. Configurations. 2009;17:105–30.
Raizada RDS, Grossberg S. Context-sensitive binding by the laminar circuits of V1 and V2: a unified model of perceptual grouping, attention, and orientation contrast. Visual Cognition. 2001;8: 431–66.
Rife DC, Snyder LH. Studies in human inheritance VI: a genetic refutation of the principles of refutation of the principles of "behavioristic" psychology. Hum Biol. 1931;3:547–59.
Rimland B. *Savant* capabilities of autistic children and their cognitive implications. In: Serban G, editor. Cognitive defects in the development of mental illness. New York: Brunner/Mazel; 1978. p. 44–63.
Rumsey J, Mannheim G, Aquino T, Gordon C, Hibbs E. Neuropsychological characteristics of calendar calculating savants with pervasive developmental disorders. Clin Neuropsychol. 1992;6:320.
Sacks O. The twins. In *the man who mistook his wife for a hat.* London: Duckworth; 1985.
Sacks O. Musicophilia: tales of music and the brain. New York, NY: Knopf Publishing Group; 2007.
Sarkar S. Einstein's universe: the challenge of dark energy. Curr Sci. 2005;89:2120–8.
Sauseng P, Klimesch W. What does phase information of oscillatory brain activity tell us about cognitive processes? Neurosci Biobehav Rev. 2008;32:1001–13.
Sherry DF, Schacter DL. The evolution of multiple memory systems. Psychol Rev. 1987;94: 439–54.
Sherry DF, Jacobs LF, Gaulin SJC. Spatial memory and adaptive specialization of the hippocampus. Trends Neurosci. 1992;15:298–303.
Shettleworth SJ. Cognition, evolution, and behavior. New York: Oxford University Press; 2009.
Singer W. Consciousness and the binding problem. Ann N Y Acad Sci. 2001;929:123–46.
Snyder A. Explaining and inducing savant skills: privileged access to lower level, less-processed information. Philos Trans R Soc Lond B Biol Sci. 2009;364:1399–405.
Snyder AW, Mitchell DJ. Is integer arithmetic fundamental to mental processing?: the mind's secret arithmetic. Proc Biol Sci. 1999;266:587–92.

Snyder AW, Mulcahy E, Taylor JL, Mitchell DJ, Sachdev P, Gandevia SC. Savant-like skills exposed in normal people by suppressing the left fronto-temporal lobe. J Integr Neurosci. 2003;2:149–58.
Spitz H. Calendar calculating, idiots savants, and *the* smart unconscious. New Ideas in Psychology. 1995;13:167–92.
Tammet D. Born on a blue day. New York, NY: Free Press; 2006.
Thorne KS. Black holes and time warps: Einstein's outrageous legacy. New York: W.W. Norton and Company; 1994.
Treffert DA. Extraordinary people: understanding. "idiot savants.". New York: Harper and Row; 1989.
Treffert DA. The savant syndrome and autistic disorder. CNS Spectr. 1999;4:57–60.
Treffert DA. Savant.com: the internet as a medical education tool. Wis Med J. 2002;101:14–6.
Treffert DA. The savant syndrome in autistic disorder. In: Casanova MF, editor. Recent developments in autism research. New York, NY: Nova Science Publishers, Inc; 2005. p. 27–55. ch. 2.
Treffert DA. The savant syndrome: an extraordinary condition. A synopsis: past, present, future. Philos Trans R Soc Lond B Biol Sci. 2009;364:1351–7.
Tuchman R, Cuccaro M, Alessandri M. Autism and epilepsy: historical perspective. Brain Dev. 2010;32:709–18.
Varela FJ, Lachaux JP, Rodriguez E, Martinerie J. The brainweb: phase synchronization and large-scale integration. Nat Rev Neurosci. 2001;2:229–39.
Viscott DS. A musical idiot *savant*. Psychiatry. 1970;33:494–515.
Walsh V, Cowey A. Transcranial magnetic stimulation and cognitive neuroscience. Nat Rev Neurosci. 2000;1:73–9.
Wass S. Distortions and disconnections: disrupted brain connectivity in autism. Brain Cogn. 2011;75:18–28.
Welling H. Prime number identification in idiot savants: can they calculate them? J Autism Dev Disord. 1994;24:199–207.
White PA. *The* structural representation of information in LTM: a possible explanation for *the* accomplishments of "idiots savants. New Ideas Psychol. 1988;6:3–14.
Whiteside DT,editor. The mathematical papers of Isaac Newton. (Volume 1, part 7 "The October 1666 Tract on Fluxions"). Cambridge University Press; 1967.
Wilson TW, Rojas DC, Reite ML, Teale PD, Rogers SJ. Children and adolescents with autism exhibit reduced MEG steady-state gamma responses. Biol Psychiatry. 2007;62:192–7.
Womelsdorf T, Fries P. Neuronal coherence during selective attentional processing and sensory-motor integration. J Physiol Paris. 2006;100:182–93.
Womelsdorf T, Fries P. The role of neuronal synchronization in selective attention. Curr Opin Neurobiol. 2007;17:154–60.
Young RL, Ridding MC, Morrell TL. Switching skills on by turning off part of the brain. Neurocase. 2004;10:215–22.
Zakharov AF, Capozziello S, De Paolis F, Ingrosso G, Nucita AA. The role of dark matter and dark energy in cosmological models: theoretical overview. Space Sci Rev. 2009;148:301–13.
Zeman A. Consciousness. Brain. 2001;124(Pt 7):1263–89.

Index

P. Bob, *Brain, Mind and Consciousness: Advances in Neuroscience Research*,
DOI 10.1007/978-1-4614-0436-1, © Springer Science+Business Media, LLC 2011

MIX
Papier aus verantwortungsvollen Quellen
Paper from responsible sources
FSC® C105338

If you have any concerns about our products,
you can contact us on
ProductSafety@springernature.com

In case Publisher is established outside the EU,
the EU authorized representative is:
Springer Nature Customer Service Center GmbH
Europaplatz 3, 69115 Heidelberg, Germany

Printed by Libri Plureos GmbH
in Hamburg, Germany